P9-CAG-558

FIREFIGHTING

FIRE
2683

FIREFIGHTING

Rob Leicester Wagner

MetroBooks

MetroBooks
An Imprint of Friedman/Fairfax Publishers

Library of Congress Cataloging-in-Publication Data

Wagner, Rob, 1954–
Firefighting / Rob Leicester Wagner.
p. cm.
Includes bibliographical references.
ISBN 1-56799-730-9
1. Fire engines. 2. Emergency vehicles. I. Title.
TH9370.W34 1999
628.9′25–dc21 99-17622
CIP

Editor: Ann Kirby
Art Director: Jeff Batzli
Designer: Howard Johnson
Photography Editor: Sarah Storey
Production Manager: Richela Fabian

Color separations by Fine Arts Repro House Co., Ltd
Printed in Hong Kong by Sing Cheong Printing Company Ltd.

1 3 5 7 9 10 8 6 4 2

For bulk purchases and special sales, please contact:
Friedman/Fairfax Publishers
Attention: Sales Department
15 West 26th Street
New York, NY 10010
212/685-6610 FAX 212/685-1307

Visit our website:
http://www.metrobooks.com

For Cory

SPEED
LIMIT
10
MPH.

Contents

Introduction

The Changing Fire Department

15 LADDER 15
15
Boston
33

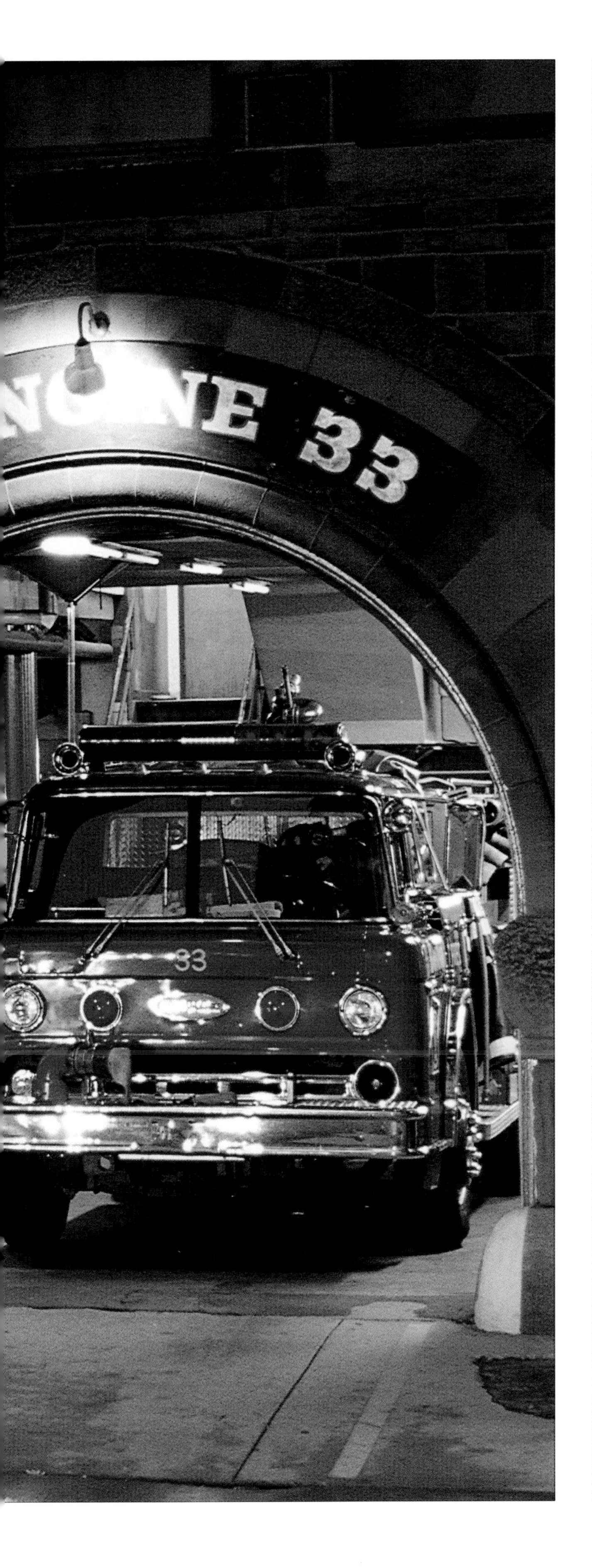

Never in American history has the work of the firefighter changed as dramatically in such a short period of time as it has in the past thirty years. In the firefighting industry, technology has often moved at a snail's pace. Following about one hundred years of ineffective battles against structure and wildfires with bucket brigades and hand-pumpers, the emergence of the steam engine revolutionized the art of firefighting in the mid-1800s. Although the revolution was a slow one, with steam apparatus virtually ignored until after the Civil War, once it took hold, the steam engine would reign supreme in the industry for about fifty years.

In the 1910s and early 1920s, the motorized age took over, and once again, fire departments were thrust into a revolutionary period of mechanized firefighting. It was a heady experience for cities and rural towns to motorize their fire apparatus, and the sentimental attachments to the old steamers made this second revolution slow in coming as well. Firefighters parted kicking and screaming with their beloved horse-drawn steamers, just as their predecessors were slow to accept steam engines.

Previous pages: *Fighting large, commercial structure fires requires near perfect coordination, not only among firefighting personnel but often with other agencies as well.* **Left:** *The Boston Fire Department houses its fire engines at night at the Boylston Street firehouse.*

Today it seems that we live in an age of constant change, when technological advancements occur at a mind-numbing pace. But unlike previous generations, the modern fire department has enthusiastically embraced new technology to fight fires and countless other emergencies now under a fire department's purview.

The experience of the Hamilton (Ontario) Fire Department typifies the changing complexity of firefighting in both Canada and the United States, which share near-identical technology and philosophy in performing their jobs.

From 1987 to 1996, reported structure fires in Hamilton dropped dramatically, from 519 calls to 458. Outdoor fires dropped even more substantially, from 802 reported in 1987 to 585 in 1996. But medical calls–including heart problems, convulsions, seizures, drownings, breathing difficulties, and strokes–went from 2,711 in 1987 to a whopping 7,666 in 1996. And that's only part of the story. Medical calls more than doubled in just a four-year period, from 1987 to 1990, when they ballooned to 6,807. Over a ten-year period, the Hamilton Fire Department averaged 6,063 medical calls, while

the ten-year average for structure fires was 508 calls, and for outdoor fires, 761.

Medium-size towns like Kalamazoo, Michigan, are faced with similar responsibilities. In 1996, 53.4 percent of the calls the Kalamazoo Fire Department responded to were for emergency medical aid. Only four of every hundred calls were for structure fires, but rescues and extrications accounted for almost 15 percent of duties.

The Kalamazoo Fire Department is well equipped to handle these calls. It has a pair of Quality 1,250-gallons-per-minute (4,731 liters per minute) pumpers; six Pierce pumpers, ranging from a mini-pumper to a 1,250 gpm (4,731lpm); a Stuphen 1,500 gpm (5,677.5lpm) with a 100-foot (30.5m) tower; and two American models, a 1,000 gpm (3,785lpm) pumper and a 75-foot (23m) ladder/pumper truck.

The much smaller city of Sierra Madre, California, is not unlike Kalamazoo, but it gets even more calls per capita for medical services, with nearly seven out of every hundred residents seeking emergency medical services, while four of every hundred calls are for structure fires. Yet because this population-ten-thousand town is nestled in the foothills, its exposure to brushfires is great. As a consequence, its inventory of apparatus may be much greater than that of another town of similar size.

The Sierra Madre Volunteer Fire Department operates three Mack pumpers–a pair of 1,500 gpm (5,677.5lpm) machines and a single 1,250 gpm (4,731lpm) model. It maintains a 1967 International brush engine that is ready for daily service, a pair of Ford rescue ambulances, a 1969 International water tender, two Chevrolet Suburbans, a 1987 GMC pickup that serves as a utility vehicle, and a 1967 Seagrave 85-foot (26m) aerial ladder. This lineup illustrates the unique position Sierra Madre is in as it fights fires on both rural and urban levels.

Not only does the expansion of a fire department's duties signify a new breed of firefighter, but the evolving strategy of how to fight fires continues to be examined. Firefighters still die in structure fires.

On January 5, 1995, four Seattle, Washington, firefighters were killed in a blaze, making it the worst fire-related disaster in the city's 109-year history. Initial crews responded to an arson fire at the Mary Pang Frozen Food Warehouse. The structure was a food processing plant and warehouse facility constructed of heavy timber.

The fire had started on the lower floor, but firefighters from the first engines on the scene entered the upper level to control the fire on the exterior as it spread to the western portion of the plant. A two-by-four framed "pony wall" had been constructed during remodeling of the plant to extend the height of the north section wall by 5 feet (1.5m). The new wall was supporting the floor joists. The floor joists were weakened by the fire and the upper level collapsed, killing the four crew members as they fell to the lower level.

Six other firefighters working the upper level escaped with their lives, but five suffered varying degrees of burns.

In all, the fire took ten hours to control and involved 144 firefighters. The Seattle Fire Department has since improved its prefire planning project, which gives fire crews full building plans accessible on a laptop computer, with clean, concise, color drawings that provide teams a clear idea of the floor plan and the hazards they may encounter.

New Technology for New Techniques

Technology that was used to build weapons during World War II and to send men into space in the late 1950s and early 1960s ultimately found its way into the products we use daily. But the plastics and chemicals that have proven so useful for

Opposite: *The pumping action of today's tankers can be hard for firefighters to get a handle on. Maintaining control of the high-powered hose requires skill and experience.* **Right:** *Firefighters spray foam during an exercise at a training facility. Foam has become an effective tool for combating a variety of blazes.*

▲ ▲ ▲ ▲ ▲

treating wood and building structures have given the firefighter another foe to battle–hazardous materials. The transportation of fuel and other hazardous chemicals on public roadways and the proliferation of clandestine drug laboratories in rural areas have also complicated the firefighter's work.

Not only must today's firefighters deal with toxic products that combust at structure, vehicle, and vegetation fires, but they must also respond to spills where chemicals can be inhaled, ingested, absorbed into the skin, and even injected. Haz-mat response teams began to appear on the firefighting scene in the 1970s and grew to full-fledged mini-departments within the authority of city and county fire agencies in a few short years.

The haz-mat operation is perhaps the first in a number of specialties within fire department ranks. Paramedic service was created as a result of the civil unrest of the 1960s, including the Watts riots of 1965 in Los Angeles and the Detroit riots of 1967. In these cases, it became apparent to the medical community that many lives could be saved if medical aid were given in the field. It became increasingly clear that gunshot and beating victims–not to mention traffic accident victims–could benefit from immediate medical attention on the scene.

The lengths to which agencies go to cope with the complex role of today's firefighter is perhaps best illustrated by the Los Angeles Fire Department. Faced with a sprawling, complicated freeway and road system, the LAFD developed a heavy rescue operation in the early 1970s. Its heavy rescue apparatus–dubbed the Heavy Utility–performs duties such as extricating victims from traffic collisions and industrial machinery entrapment, moving debris from collapsed structures (as witnessed in the 1994 Northridge earthquake), and general heavy lifting.

Assigned to Fire Station 56 on Rowena Avenue, LAFD's Heavy Utility is a 1995 Peterbilt 377 A/E conventional tractor rig with a Century wrecker body. It's powered by a Cummins diesel engine that generates 460 horsepower at 1700 rpm. It's also equipped with an Eaton-Fuller transmission with eighteen forward and four reverse gears. It has a boom capacity of 80,000 pounds (36,320kg) and a winch motor capacity of 45,000 pounds (20,430kg). The breaking strength for its cables is a stiff 49,700 pounds (22,564kg).

In British Columbia, Vancouver Fire and Rescue Services has made dramatic

Waterfront fires like this one in Oakland, California, require special equipment and training.

ABOVE: *Extreme weather presents extreme challenges to firefighters responding to emergencies. Here, the Patterson (New Jersey) Fire Department battles a blaze in the depths of winter. Falling snow and water from the hoses have caused the truck and the ladder to ice over in the below-freezing temperatures.* **OPPOSITE:** *Owosso firefighters prepare to fight a small blaze .*

▲ ▲ ▲ ▲ ▲ ▲ ▲ ▲ ▲ ▲ ▲ ▲ ▲ ▲ ▲

changes in its operations during the 1990s to meet the changing needs of fire and rescue operations. By the latter part of the decade, Vancouver had 792 personnel assigned to nineteen fire halls. The department is equipped with nineteen fire engines, fourteen aerial units, two rescue units, a hazardous response vehicle, and five fireboats. In 1993, the department responded to 36,853 incidents, including 2,677 fire calls and 25,427 first-responder emergency medical calls. As a result of the increasing medical calls, the department started a program in the mid-1990s to train firefighters in using automatic external defibrillators. Every front-line apparatus is now equipped with the defibrillators.

In 1995, Vancouver also began HUSAR, its Heavy Urban Search and Rescue Team program, to handle the unusually dirty and tricky jobs of large-scale search and rescue efforts in congested urban areas.

Rescue operations, once limited to the military, are now commonplace among fire departments, thanks in large part to rescue techniques learned during the Vietnam War. Fire department helicopters began to appear on the scene in the late 1950s and early 1960s, after the Korean War proved that choppers could be effective in rescue and scouting operations. The Vietnam War helped refine these operations and showed that complicated rescues could be performed that could save hundreds of lives in a single day. Today's city, state, and federal fire agencies are equipped with extensive fleets of choppers–including models by Bell, Boeing, and Agusta–to scout and douse wildland fires and perform risky rescue operations.

Complementing the choppers are the massive air tankers–commonly DC-4s

and PBYs. Able to take on a load of fire retardant in less than ten minutes at an air base, they are used to douse fires in terrain impossible to cover from the ground.

On the water, fire departments in the cities of Los Angeles, Seattle, and New York, among others, maintain fleets in varying numbers of fireboats and swift-water rescue vessels to fight fires and provide rescues to distressed craft, boaters, and swimmers. Training swift-water rescue teams is an expensive necessity; in Los Angeles, for example, the LAFD spends $3,757 to train each swift-water team member, bringing the total annual cost for its forty-eight-member team to $180,336.

Firefighters' new responsibilities–including haz-mat response, rescue, and paramedic operations–as well as heavier equipment to get a better edge on wild-land fires and airport emergencies, evolved in less than twenty years. As a result, firefighting has become highly specialized, and is performed by men and women better trained than firefighters of any previous generation.

With these new responsibilities and strategies has come a new breed of custom and commercial apparatus builders. Replacing such venerable names as Mack, Pirsch & Sons, Seagrave, Crown, and American LaFrance are Emergency One (E-One), the Kovatch Corp. (KME), Pierce, and smaller entities like Marion Body Works Inc., Alexis Fire Equipment, and Luverne Fire Apparatus.

Out with the Old, In with the New

E-One and KME have in recent years emerged as leaders in firefighting equipment. If these companies signal the dawn of a new era in firefighting technology, then Crown typifies the passing of an era, although it helped pioneer modern efforts. Akin to the phasing out of the legendary Ahrens-Fox fire apparatus–or better yet, because of its regional influence, the Buffalo Fire Engine of New York in the 1950s–Crown slowly faded away, but not without a few tears from its loyal operators.

Founded in the early 1950s, Crown Coach Corp. of Los Angeles at one time rivaled the big builders of fire engines in the Los Angeles area.

A testament to Crown products' rugged durability is the fact that the LAFD took delivery of its last Crown apparatus in 1973, but Crowns continued to see service in various roles through the late 1990s. By the mid-1970s, Crown had ceased production altogether, but many firefighters today recall hearing the distinct sound of Crown's gasoline-powered Hall-Scott engines, and using its Waterous pumps.

The first Crown delivered to the LAFD was a 1953 triple-combination pumper with

a 1,250 gpm (4,731lpm) pump and 350-gallon (1,325lpm) water tank. It was powered by a 953-cubic-inch Hall-Scott engine.

Because of their reliability, Crown fire coaches saw service in many different incarnations, often achieving second careers. The LAFD once used a Crown as a foundation, cannibalizing three different rigs to develop a hybrid truck that evolved into a Crown-Seagrave aerial ladder truck.

Some Crowns can still be seen on southern California roads, but most have been mothballed. At the Los Angeles County Fire Department Headquarters on Eastern Avenue, three Crown pumpers stand as silent sentinels, their paint faded and chrome dulled, but still appearing ready for action.

Still, it's fitting for fire engine makers to pass the torch as fire departments enter a new century.

Major chassis builders that once took a back seat in providing firefighting equipment, like Ford and Freightliner, are now leaders. Freightliner in particular, which has roots in long-haul tractor trailer rigs, has emerged as a leader in providing chassis and cabs for airport tenders, wildland tankers and pumpers, and industrial fire engines.

During the 1990s, Emergency One (E-One) became one of the leaders in state-of-the-art firefighting technology with its line of pumpers and tankers, aerials, airport

Opposite: *Air tankers dropping fire retardant and water have become perhaps the most effective means of battling difficult wildfires in areas where conventional firefighting equipment can't go. Using air tankers, however, is dangerous and expensive.* **Above:** *Firefighters operate a deck gun from an engine during a nighttime operation.*

machines, industrial vehicles, and rescue and wildland vehicles. The Ocala, Florida-based E-One was founded in 1974 by Robert Wormser, who operated his business out of his barn. Specializing in an all-aluminum fire truck body, his business quickly grew. It was bought by the Federal Signal Corp. of Elgin, Illinois, in 1979. By the mid-1990s, it had become an international builder of fire rescue vehicles with more than sixteen thousand vehicles operating worldwide.

Today, E-One boasts more than sixteen hundred employees working in five separate plants that total more than 500,000 square feet (46,500 sq m). It specializes in building complete fire engines with chassis, cab, body, tank, and aerial devices. Its products average about $100,000 per unit and include warranties, training sessions, and manuals for fire departments.

KME began as the Kovatch Corp. in 1946 and is now based in Nesquehoning, Pennsylvania. It was first founded as an auto repair shop by John Kovatch, Sr., and his sons John Jr. and Joe. The company grew to include automobile sales and services before changing to manufacturing in the mid-1970s. It incorporated KME Fire Apparatus in 1985, and in less than a decade it rivaled E-One as the leader in fire equipment manufacturing and sales. Before building fire equipment, KME had earned a strong reputation as a quality vehicle builder with extensive contracts to build vehicles and aircraft refuelers for the

U.S. Air Force, Army, Navy, Marines, and Coast Guard. In addition to manufacturing its current crop of fire apparatus, it remains the builder of choice for military refuelers.

Kovatch's Nesquehoning plant covers 65 acres (26ha) with more than 500,000 square feet (46,500 sq m) of manufacturing and equipment space. KME employees number 875, with plants in Roanoke, Virginia, and Ontario, California. Annual output for the builder is four hundred trucks, and $100 million in sales. Like E-One, KME builds fire engines from the rails up, with its own chassis, bodies, cabs, tanks, and even ladders.

Pierce Manufacturing Inc. is another industry leader in custom fire and commercial apparatus. A subsidiary of Oshkosh Truck Corp., Pierce is based in Appleton, Wisconsin. The company features a long line of custom and commercial pumpers, aerial units, rescue and wildland trucks, mini-pumpers, and elliptical tankers.

In May 1997, Pierce took an order valued at $7.5 million from the Kern County (California) Fire Department for twenty-eight fire trucks to replace its aging fleet. The agency serves an 8,000-square-mile (20,720 sq km) area that includes considerable wildland. Pierce also took a $35 million order in 1997 from the Saudi Arabia Civil Defense for 130 fire trucks, with delivery expected in November 1999.

Smaller companies like Marion Body Works offer important alternatives to the major builders. An independent and - privately held manufacturer in Marion, Wisconsin, the company was founded in 1905. Its plant covers 52 acres (21ha) with 115,000 square feet (10.5 sq m) of manufacturing space. In recent years, it has established a niche in building specialized trucks and transportation equipment, with an emphasis on all-aluminum bodies for emergency vehicles. It employs about 130 workers, and its annual sales range from $12 million to $15 million.

Another growing equipment company is Alexis Fire Equipment. It was founded in 1947 and specializes in custom apparatus, including its Vision 2000 Primary Response Unit, which has an aluminum body and shorter turning radius than any two-wheel-drive pickup truck.

A more established company, with nearly a century of firefighting equipment experience, is Luverne Fire Apparatus. Founded in 1896 as Luverne Wagon Works in Luverne, Minnesota, it began building fire engine pumpers in 1912. In 1985, it moved to a new plant in Brandon, South Dakota.

The public today often takes for granted the sophistication and progress of firefighting technology. But, due in part to newly developed equipment that allows firefighters to perform more complex tasks with less risk to life and limb, loss of life due to fires is no longer measured in the scores and hundreds but in fractions of former numbers. From wildfires in the hills above Los Angeles to terrorist bombings of buildings, the death toll has been remarkably low, thanks to state-of-the-art technology engineered by front-line firefighters and administrators with a vision.

Above, left: *Fireboats have been serving waterfront communities for about a century. Like land machines, these vessels have a strong following among firefighters and firefighting-equipment enthusiasts.* **Opposite:** *Battling odds that usually favor nature, the firefighter must exhibit courage and strength in the face of life-threatening challenges.*

▲ ▲ ▲ ▲ ▲

ENGINE TANK
2-28
AMERICAN LAFRANCE
COLCHESTER
HAYWARD
F.D.

1

Pumpers and Tankers

Previous pages: *A 1982 American LaFrance pumper stands at the ready for Company 2 of the Colchester-Hayward (Connecticut) Fire Department. While Emergency One and other relatively new fire engines dominate the market, the American LaFrance is a perennial favorite among veteran firefighters.* **Opposite:** *A lone firefighter tends to his pumper during an early morning blaze.* **Above:** *The Northside (California) Fire Department is perfectly content using its 1965 Crown 1,500 gpm (5,678lpm) pumper, which continues to be just as reliable as it was the day it rolled off the assembly line. This is one of only about ten three-axle Crown pumpers built.*

Considered the worst fire in California history, the Oakland fire of October 1991 demonstrated the fickle nature and unpredictability of fire and its terrible consequences.

On Sunday, October 20, the Oakland Fire Department responded to a blaze in the hills of Oakland, an upper-middle-class section of town with two-story homes, well-manicured lawns, and swimming pools. By early afternoon, the blaze was out of control and mutual aid from neighboring cities was desperately needed.

Through the State of California Mutual Aid Plan and the Bay Area Inter-County Mutual Aid Plan, the city of San Francisco and other nearby agencies responded to Oakland's pleas. Fresh firefighters quickly arrived on the scene. Their first impression was that the blaze had all the characteristics of a firestorm.

"I saw embers flying–I actually saw little balls of fire flying all around in the air," recalled San Francisco Fire Lt. King A. Strong, an eighteen-year veteran of the department. "You'd see fire flying everywhere. The intensity of it was incredible. I heard of firestorms, but this was the first time I was ever in one."

Beacons atop engines melted. Engineers cut their way through downed wires to "get to neighborhoods and had to wrestle with 45-degree grades to get to hilltop areas. Civilians chipped in by pulling hose, and firefighters were grateful for the help. Whole neighborhoods were consumed by flames, which leapt from house to house. Occasionally–and inexplicably–a single wood-frame house would be spared while every other structure on a street was consumed.

"Everything along the ridge line was on fire," said one firefighter. "All the buildings were going pretty good. We just picked a building and saved it. The fire went past us and over us. The wind speeds were up to 70 miles per hour [112.5kph]. Fire was blowing everywhere."

Virtually every firefighter from Oakland, San Francisco, and Berkeley responded to the disaster. Hundreds of calls flooded fire department switchboards as smoke and ash covered the Bay Area.

In all, 1,600 acres (640ha) burned and twenty-nine hundred structures were damaged or destroyed. Twenty-five people died.

The Heart of Firefighting

Pumpers and tankers, the backbone of daily firefighting, called to respond to virtually all calls whether medical aid or rescue operation, were one reason the death toll and structure damage were not greater during the Oakland fire. In hard-to-reach locations, these pumpers provided enough water to control certain areas of the conflagration.

Commercial appliances see heavy duty in small towns and cities on a tight budget. The so-called off-the-rack fire engines were often built in the early motorized days of firefighting on Ford, Chevrolet, and General Motors chassis. A fire equipment builder would then add racks for ladders, sirens, and compartments for tools, hoses, and pumps. Even now, these engines are less expensive and are limited in features. Custom units are tailored specifically to the customer's demands and are generally reserved for more wealthy communities.

Ford and Freightliner, among a handful of other truck makers, are perhaps today's leaders in providing chassis for commercial pumpers and tankers.

Ford

Until the last decade, Ford had never been a major player in big-rig trucking, preferring to focus its attention on the light-duty truck market. There were exceptions, of course, but its production numbers were never particularly high with big rigs.

Ford developed a 2-ton (1.8t) cab-over-engine model in 1926 under its Fordson custom-built line. And its revolutionary V8 engine, produced in 1932, gave the truck maker standing in the 1930s by powering its new tractor rigs, which featured sleeper cabs.

Still, Ford was hampered by its reluctance to enter the diesel engine field. While its larger F-Series trucks were impressive, long hauls and heavy work were not part of the Ford vision. That changed in 1959, when Ford finally introduced its first diesel.

In 1969, Ford took another big step in the big-rig field by opening its truck plant in Louisville, Kentucky, to focus on building tractor trailer rigs. The CL 9000

This 1997 Freightliner is fitted with a high-pressure pump, a 1,000-gallon (3,785l) water tank, and 50-gallon (189l) foam tank capability. The fire truck serves the Merced (California) Fire Department.

cab-over-engine model was introduced in 1977 and ran through 1991, establishing Ford as a leader in the industry.

Emergency One

Emergency One (E-One) has emerged as one of the top makers of commercial and custom pumpers and tankers in the last fifteen years, eclipsing old standbys such as Mack, Crown, and American LaFrance.

The Knoxville (Tennessee) Fire Department has long been a customer of E-One, relying heavily on the manufacturer to supply its fleet of pumpers. Protecting more than 200,000 people and responding to more than twenty-six thousand calls in 1996 alone, Knoxville operates twenty engine and ladder companies, two mini-pumper companies, a rescue company, and a haz-mat team. It operates twenty-nine E-One Hurricane engines, twenty-one of which are pumper/tankers.

E-One frequently uses both Ford and Freightliner chassis for its commercial pumpers and tankers. Its newest offering

A Ford Super Tanker serves the Owosso Township (Missouri) Fire Department. In recent years, Ford has emerged as one of the leading fire engine manufacturers.

in the commercial category is a pumper placed on a Ford F-800 chassis with a 35,000-pound (15,890kg) gross vehicle weight rating and a 190-inch (482.5cm) wheelbase. Powered by a Ford diesel 250-horsepower engine with an Allison MD 3060P automatic transmission, the pumper is state-of-the-art.

Part of a line dubbed the American Eagle series, the pumpers are equipped with steel disc wheels, a 50-gallon (189l) step tank, power steering, and a heavy-duty cooling system.

The American Eagle body consists of a 1,250 gpm (4,731lpm) single-stage pump with suction relief valve, a stainless steel pump panel and color-coded pump panel tags, a 1,000-gallon (3,785l) polypropylene tank, and 55 cubic feet (1.5 cu m) of hose bed with a single adjustable divider.

Particularly important for extensive use on major alarms are the large compartment storage areas, which are vented and lighted. The doors on the truck are equipped with gas shock door holders. The truck is also equipped with a Federal PA 300 siren and a 100-watt siren speaker.

E-One's American Eagle tanker is fitted on an identical Ford F-800 chassis, whose body is equipped with a 300 gpm

(1,135.5lpm) pump; stainless steel, brass, or flex hose for all plumbing; an 1,800-gallon (6,813l) polypropylene tank; a water-level gauge with noncorrosive sender; and a fold-down, portable tank rack on the right side.

Another pumper in E-One's commercial line is the Freightliner FL80, a conventional four-door rig placed on a 242-inch (614.5cm) wheelbase and with a gross vehicle weight of 40,000 pounds (18,160kg). Freightliner has been a fixture of big-rig trucking since Leland James founded Freightways Inc. in 1939. Freightways became Freightliner in 1947, the year a new plant was built in Portland, Oregon. The company was a leader in the extensive use of aluminum and magnesium, allowing its trucks to weigh as much as 1 ton (908kg) less than competing tractor trailer rigs on the road.

In addition to its focus on lightweight construction, Freightliner pioneered the use of interchangeable parts in the 1970s. During that decade, it also began featuring the Cummins NTC-90 diesel engine. The company was purchased by Daimler-Benz in 1981, and by 1992 it accounted for nearly a quarter million tractor trailer rigs on U.S. roadways.

E-One's Freightliner commercial pumper is powered by a Cummins C8.5 300-horsepower engine with an Allison MD-3060P five-speed automatic transmission. Its pumper sports a 75-gallon (284l) fuel tank, air brakes with automatic slack adjusters, and steel disc wheels.

The cab demonstrates just how far firefighter safety and comfort have come since the days of driving open trucks and fire engines in inclement weather. This Freightliner four-door crew cab measures 153 inches (388.5cm) across with seating for five. The cab features tinted glass, heat, air conditioning, exterior grab handles, and dual padded interior sun visors.

The body is all aluminum, with sidewall thickness measured at 3/17 inch (4.5mm) to give it solid strength but still keep the weight down. There are four extruded-aluminum hard-suction-hose racks, plus a hose bed compartment deck that is constructed of anodized hollow aluminum extrusions that gives the hose bed 84 cubic feet (2.3 cu m) in capacity. A single ladder is mounted on the right-side pump module to provide access to the deck gun.

Equipment on the Freightliner includes a 1,000-gallon (3,785l) foam tank, which is a 1/2-inch (12mm) polypropylene shell with 3/8-inch (9.5mm) baffles and a 20-inch

Opposite: *This 1995 Emergency One pumper "Hurricane" offers a 2,000 gpm (7,570lpm) pump with foam capabilities. It belongs to the fire department in Everett, Washington.* **Above:** *The Dunklin (Missouri) Fire Department uses a Chevrolet/Pierce combination for light duty. The fire truck is equipped with a 300 gpm (1,136lpm) pump.*

▲ ▲ ▲ ▲ ▲

The Woodinville (Washington) Fire Department uses a 1996 HME/Smeal 750 gpm (2,839lpm) pump, 1,500-gallon (5,678l) unit with foam capabilities. This unit comes equipped with bicycles for aid calls on local bike paths.

(51cm) hatch. The foam system is a Williams Hot Shot II 150, balanced pressure rated at 2,500 gpm (9,462.5lpm). The deck gun is a Williams Patriot I HydroChem 1,000 gpm (3,785lpm) with a tiller-style handle control.

Mounted midship is a 1,500-gpm (5,677.5lpm) centrifugal pump. Plumbing includes a pair of 6-inch steamer inlets, four 2½-inch discharges, and a single right-side discharge with a handwheel control on the pump panel. The chassis and frame rails are painted black, and the pumper itself is painted lime yellow, red, white, or chrome yellow.

Specifications for E-One's custom pumpers can differ widely from the commercial offerings, but a typical custom unit features an Emergency One Cyclone II Cab on a 225-inch (571.5cm) wheelbase with a massive gross vehicle weight rated at 51,000 pounds (23,154kg). It is powered by a Detroit Diesel Series 60 500-horsepower engine with an Allison HD-4060P five-speed automatic transmission. Its fuel tank is very large at 100 gallons (378.5l).

The unit seats six firefighters in a custom four-door tilt cab with a fully padded interior. The tilt is effected by a hydraulic pump with manual override to power the tilt cylinder. Another feature found on the custom cab is three-point seatbelts for all six crew members.

Total hose bed capacity is 84 cubic feet (2.3 cu m), with a minimum of 118 cubic feet (3.3 cu m) for total storage space that includes enclosed compartments for reels, hose lines, and ancillary firefighting equipment.

The foam tank for this unit has a 1,000-gallon (3,785l) capacity; the capacity of the Williams Hot Shot II is 4,000 gallons (15,140l); and the capacity of the Williams Patriot II HydroChem is 2,000 gallons (7,570l). Like the Freightliner version, the custom pumper has a single-stage centrifugal pump mounted midship, but the gpm rating is considerably higher, at 3,500 gpm (13,247.5lpm).

The pump panel, which can be removed or hinged, is brushed stainless steel. Its discharges and intakes are color-coded. Mounted on the panel are the pump operator panel lights; gauges for the master drain valve, electric primer, engine cooler, foam tank level, fuel, oil pressure, and water temperature; and an oil pressure and water temperature alarm.

In 1997, the Center Township in Marion, Indiana, purchased an E-One Cyclone II rescue pumper for its twenty-two-member volunteer fire department. Covering about 35 square miles (90.5 sq km), the fire department uses four trucks, two Hendrickson engines, and a Hendrickson tanker, along with a 1-ton (908kg) 1996 Dodge dualie with utility box to serve as the department's squad unit.

Marion volunteer firefighter Bryon Williams said that Chief George Smith and his staff designed the pumper to their specifications. They needed a rescue truck to carry large equipment loads and to serve as their front-line pumper. Smith and his staff requested a customized 1,000-gallon (3,785l) tank plus compartments capable of storing hydraulic rescue tools. Marion's E-One pumper is used for a full range of services, averaging five to ten calls per week. It is used for extrications from crushed vehicles and for structure fires.

In addition to using Freightliner and Ford chassis for its commercial line, E-One offers chassis from Navistar, GMC/White, Volvo, Peterbilt, and Kenworth. As many as nine different commercial pumpers are offered on these chassis, ranging from the 190-inch-wheelbase (482.5cm) two-door side-mount pumpers to the 250-inch-wheelbase (635cm) four-door top-mount pumpers.

Kovatch Corp. (KME)

While KME provides commercial chassis as E-One does, its custom Excel chassis puts the company squarely in the lead in state-of-the-art technology with a number of cab configurations, including an optional 16- or 22-inch (40.5 or 56cm) raised roof or the standard 6-inch (15cm) rear raised roof. Seating for these cabs ranges from five to ten passengers. KME constructs its own cabs, virtually all of which are manu-

The Love (Mississippi) Fire Department takes on blazes with a 1989 Freightliner/95 Gilmans tanker truck possessing a 500 gpm (1,893lpm) pump and a 3,000-gallon (11,356l) water tank.

factured to customer specifications with no restrictions. These cabs also feature marine-grade gauges with integral indicator lights and central warning light clusters, as well as custom-formed overhead consoles, formed door panels with integral storage pockets, and engine enclosures and compartments for quick access.

Perhaps the best example of these virtual luxury cabs is the KME Renegade LFD 1,750 gpm (6,624lpm) custom pumper. Equipped with a 75-foot (23m) Firestix, this raised-roof cab tilts for access to the engine and can seat eight firefighters. It sits on a 192-inch (487.5cm) wheelbase, with a massive overall length of 406 inches (1,031cm).

It's also equipped with a 500-gallon (189l) polypropylene fiberglass tank, a three-way communication system, and a three-section steel ladder with 12-inch (30.5cm) handrails.

A smaller version is the Renegade MFD 1,250 gpm (4,731lpm) top-mounted pumper. The custom cab is a tilt, five-man model with an 8-inch (20.5cm) raised roof over the crew seating area. It sits on a wheelbase of 184 inches (467cm) with an overall length of 351 inches (891.5cm). It contains a top-mount operator's panel, a 1,250 gpm (4,731lpm) single-stage pump, and a 3-inch (7.5cm) deck gun discharge.

The fire department in Placer Foot Hills, California, took delivery of a KME pumper featuring an International 4900 commercial cab and chassis with a 1,250 gpm (4,731lpm) Hale pump and 750-

This 1991 Pierce ET fire engine is owned by the volunteer fire company in Salem, Oregon. Pierce Manufacturing Inc. is a leader in the industry with custom fire and commercial apparatus. A subsidiary of Oshkosh Truck Corp., Pierce offers custom and commercial pumpers, aerial units, rescue and wildland trucks, mini-pumpers, and elliptical tankers.

This 1995 Spartan Gladiator-Anderson fire engine is owned by the Colwood Fire-Rescue Department in British Columbia, Canada. It has a 1,500-gallon (5,678l) water tank and a 500 gpm (1,893lpm) pump with a foam tank as well.

gallon (2,839l) water tank. The pumper is powered by an International DT 466 300-horsepower engine with an Allison MD 3060 automatic transmission.

Another version of this type of pumper is the Freightliner FL80 commercial cab and chassis with a KME rear-mount pumper body, Cummins C-Series 300-horsepower diesel engine, Allison MD 3060 P automatic transmission, 1,000-gallon (3,785l) water tank, and Hale pump.

The Air Force Academy Fire Department in Colorado Springs, Colorado, has four KME pumpers similar to the Freightliner models. Three of the academy's pumper fleet are KME P-22 models with 1,000 gpm (3,785lpm) and 600-gallon (2,271l) tanks. The fourth pumper is a P-24, a four-wheel-drive version with pump and roll capability and a heavy-duty winch. These four models are equipped with 55-gallon (208l) foam tanks.

The KME tandem tanker/pumper is similar to some of the tanker/pumper machines that saw service in the treacherous Oakland blaze. KME's version is a two-door tandem commercial chassis on a 214-inch (543.5cm) wheelbase, featuring a 1,500 gpm (5,677.5lpm) single-stage pump, a pair of 6-inch (15cm) main suction inlets, one 2 1/2-inch (6.5cm) auxiliary suction, six 2 1/2-inch (6.5cm) discharges, and a 3-inch (7.5cm) deck gun discharge nestled directly behind the cab. There's an electrically actuated 10-inch (25.5cm) square dump valve at the rear and a series of roomy compartments on the left side, with pullout steps in the lower compartments.

24

2

Aerials

Previous pages: *An aerial and platform aerial battle a fire. It takes precision and skill to effectively fight fires from above. Aerials are also the most important piece of equipment when fighting high-rise fires in congested areas.* **Above:** *Two firefighters spray water from a crow's nest into a commercial blaze. A task force team of firefighters uses three pieces of apparatus for combating fires in tall office buildings: the aerial truck, an engine company, and a single pump engine. The truck crew usually consists of a captain, an apparatus operator, and three firefighters.* **Right:** *Note the height advantage these firefighters have over a two-story structure. Their position gives them not only an effective angle for spraying water onto a concentrated area, but also a bird's-eye view of the entire scene.*

▲ ▲ ▲ ▲ ▲

The city of Los Angeles has weathered more disasters–natural and man-made– and more civil unrest in the 1990s than any other major metropolitan area in the United States.

The city's unusual proximity to the Pacific Ocean, mountain ranges, and deep valleys, coupled with its high-rise structures downtown, has created unique challenges for the city's fire department.

A burgeoning population, and its resulting conflicts with nature, began to manifest after World War II, but the harbinger of what the city and county of Los Angeles faced in complicated, detailed firefighting came with the Bel Air fire in November 1961. It ranks as California's fifth worst blaze, and as the worst ever in southern California.

The fire swept through the posh neighborhoods and hilly terrain for two days, destroying more than five hundred structures and consuming an estimated 6,090 acres (2,436ha) of valuable watershed. Firefighters had to contend with the infamous Santa Ana winds, which reached up to 50 mph (80.5kph) as they carried thousands of burning brands to other areas. Not a single life was lost, but the destruction was enough to prompt legislation regulating building materials throughout the state and brush conditions in the mountain areas.

Los Angeles' city and county fire departments began to rethink the type of equipment they used. At the same time, a city ordinance that had been enacted in 1928, which restricted any building from being taller than City Hall, was repealed. It allowed skyscrapers to come to Los Angeles. The first high-rise building–the Occidental Tower, now the TransAmerica Building–was constructed in 1962. The aerial subsequently took center stage as one of the most important firefighting tools in the city.

Left: *Fires in older multi-unit structures such as this one are particularly dangerous and difficult to combat, making an aerial platform essential.* **Above:** *A Pierce Tele-Squirt at the site of a live-fire training exercise in 1997.*

▲ ▲ ▲ ▲ ▲

Los Angeles' aerial trucks are the province of the fire department's task force, a team of firefighters that uses three pieces of apparatus: the aerial truck, an engine company, and a single-pump engine. The truck crew consists of a captain, an apparatus operator, and three firefighters.

Firefighters characterize the aerial truck as a hardware store on wheels. It carries ladders ranging from 12 to 100 feet (3.5 to 30.5m). It contains tools for forcible entry into buildings, ventilating structures, or performing salvage work and aboveground rescue.

The Encinitas (California) Fire Department uses a state-of-the-art 1996 Spartan-LTI with a 1,500-gallon (5,678l) water tank and a 100-foot (30.5m) aerial.

Aerials in Action

Emergency One has supplied about twelve hundred aerials to fire departments throughout the world. Most are constructed on commercial chassis, including Ford, GMC, Navistar, Freightliner, Mack, Peterbilt, and Kenworth. Chassis options include the Hush rear-engine model, Hurricane, Tilt, and Cyclone II. The aerial's midmounted waterway dual base measures 4 inches (10cm), with the dual midsection waterway measuring 3 1/2 inches (7.5cm).

The midmounted boom gives firefighters easy access to the hose bed, enhancing deployment of attack lines. It also allows for improved weight distribution, giving the truck better maneuverability and stability. Each boom provides optional controls on the right side of the mid-mounted aerial. Rear-mounted booms and their optional boom controls are found on the pump panel and allow a single firefighter to control both the pump and the boom.

Each aerial truck is equipped with an outrigger jack spread, which is very narrow, spreading less than 11 inches (28cm) from the body of the truck, to allow it to be set up in narrow alleys and congested areas. The spread allows vehicle stability.

E-One's custom Quint aerials typify the high-tech aerials now operated by fire departments. Two custom rear-mounted boom models for the smaller aerials are offered: the HP 75 (the ladder extends 75 feet [23m]) Hurricane fixed-cab aerial and the HP 75 Cyclone II tilt-cab aerial. Wheelbase for these models ranges from 200 to 230 inches (508 to 584cm); they are powered by Cummins, Detroit Diesel, or Mack engines with horsepower ratings of 350 to 500. The outrigger is a dual H-style underslung out and down with a single switch control for each jack leg and a spread of 16 feet (5m) from pin to pin. Its rated tip capacity is 500 pounds (227kg) wet or dry.

The aerial controls are ergonomically designed to provide a light touch to give the operator a better sense of control over the boom. The aluminum ladder is wide with high handrails to provide enough room for firefighters to move around with

A 1989 Pacific with a massive 135-foot (41.1m) Bronto Skylift serves the Ottawa (Ontario, Canada) Fire Department. The Bronto telescopic can withstand 31 mph (50kph) winds. It comes equipped with such options as a 110-volt power supply, quartz lights with telescopic poles, and collision guards. The boom swings 180 degrees and allows firefighters to reach over large obstacles and gain access to lower roofs. It can operate 15 feet x(4.6m) below grade and provides a rescue ladder on the right.

▲ ▲ ▲ ▲ ▲ ▲ ▲ ▲ ▲ ▲

equipment. The rear body is designed to allow easy access to the turntable.

The Tampa (Florida) Fire Rescue Department took delivery in 1997 of an 95-foot (29m) E-One platform designed specifically for work in heavily congested areas. Tampa's aerial is powered by a 470-horsepower Series 60 Detroit Diesel engine with seating for six in a Hurricane cab. E-One customized Tampa's platform to provide 189 1/2 cubic feet (5.3 cu m) of storage space. Platform rating is 1,025 pounds (465kg), with wide ladder access to allow two firefighters to pass on high-rise evacuations.

The CR 100 aerial by E-One is a different animal altogether. Its Hurricane tilt four-door cab seats up to seven firefighters with a low profile of just 96 inches (244cm) from ground to roof. Wheelbase ranges from 220 to 252 inches (559 to 640cm). Power comes from a Cummins or a Detroit Diesel, with horsepower ranging from 400 to 470. The 100-foot (30.5m) aerial has a rated 750-pound (340.5kg) tip load, with 200- to 500-gallon (757 to 1,892.5l) polypropylene tanks available. Another aerial, the HP 105 platform, sits on a 250-inch (635cm) wheelbase without a pump, 258 inches (655cm) with a pump. It's powered by a Detroit Diesel, Cummins, or Mack E7 diesel engine. The platform is all aluminum with an integral heat shield and shower curtain nozzle.

Perhaps the most versatile aerial is E-One's Heavy Duty Telescopic 2000. The Bronto telescopic offers a 20-square-foot (52 sq km) platform area that can withstand 31 mph (50kph) winds, a fold-down railing, and a 1,000-pound (454kg) lifting eye. The stainless steel waterway pipes are side-mounted on the boom to avoid contact with buildings. The aerial also comes equipped with options like a 110-volt power supply, quartz lights with telescopic poles, and collision guards. The boom swings 180 degrees and allows firefighters to reach over large obstacles to place personnel on lower roofs. It can operate 15 feet (4.5m) below grade and provides a rescue ladder on the right side.

The Ottawa (Ontario) Fire Department uses a similar Bronto tower ladder to handle high-structure work. Assembled by Anderson Engineering of British Columbia with a 65,000-pound (29,510kg) Pacific chassis, Ottawa's tower-ladder fire engine is powered by a 500-horsepower turbocharged Detroit Diesel engine. Its

Bronto skylift is a 107-foot (405m) elevating platform.

The Aerialcat platform manufactured by KME has a 1,000-pound (454kg) rating at full extension and can be angled from –5 degrees to 80 degrees. It contains a 2,000 gpm (7,570l) water flow from its dual deck-gun monitors and is rear-mounted to fit into a 12-foot (3.5m) stationhouse door. An automatic leveling system for the outriggers is operated with a single switch. The truck can level itself in less than thirty seconds.

KME's Aerialcat also comes in a 95-foot (29m) midmount platform that allows it to fit into a 10-foot (3m) stationhouse door. The company's Firestix version comes in a 55- or 75-foot (17 or 23m) version with 12-inch (30.5cm) steel handrails on the rescue ladder, rear and pump panel aerial control stations, a communications system from the tip to the rear control console and pump panel, and a single set of outriggers. The engine is loaded with 500 gallons (1,892.5l) of

Kansas City, Missouri, firefighters use a 100-foot (30.5m) aerial to battle a blaze at a construction site. Aerials became a necessity in some cities when officials lifted bans on high-rise structures. The city of Los Angeles, for example, began allowing skyscrapers in 1962, resulting in a need for aerials.

▲ ▲ ▲ ▲ ▲

Left: *The Shaker Heights (Ohio) Fire Department uses a Simon-Duplex aerial. Note the outrigger jack spread that stabilizes the truck and aerial while in use.* **Above:** *The North Cowichan (British Columbia, Canada) Fire Department uses this American-made 1996 Emergency One with a 65-foot (19.8m) Quint aerial.*

▲ ▲ ▲ ▲ ▲

water and a 500-pound (227kg) tip load from 0 to 75 degrees. Its 1,000 gpm (3,785l) water tower can function from –10 to 90 degrees.

While E-One, KME, and many other large builders of fire apparatus manufacture a wide range of firefighting and emergency equipment, other companies are more specialized. Aerialscope Inc., a Richmond, Virginia-based company, focuses its attention solely on the construction of aerials.

A New York City Fire Department hook-and-ladder fire truck on a residential street.

▲ ▲ ▲ ▲ ▲ ▲ ▲ ▲ ▲ ▲ ▲ ▲ ▲ ▲ ▲

Aerialscope's roots date back to the early 1960s, when the New York City Fire Department sought a more versatile aerial that could be effectively deployed on narrow city streets and in confined areas. At the time, New York's largest supplier of engines was Mack Trucks Inc., a company that supplied a large percentage of the United States' fire apparatus (and which continued to do so through the 1980s). Mack began a relationship with Eaton Metal Products in Colorado to manufacture a telescoping crane that would feature a platform at the end of the boom and a waterway to carry water to the platform. The project never got off the ground, but the product design and patent were sold to Mack.

By July 1969, Mack joined with Baker Equipment of Richmond to reexamine the design. The result was the Mack 75 Aerialscope. The Bedford (Ohio) Fire Department took delivery of one in late 1970, and within fifteen years more than three hundred such units were sold worldwide. By 1986, a 95-foot (29m) aerial was completed and delivered to the Providence (Rhode Island) Fire Department.

Baker and Mack continued their relationship. In January 1988, the two companies signed an agreement that allowed Baker to market the Aerialscope under the Baker name. When Mack stopped production of its fire apparatus chassis in March 1990, the Aerialscope began to appear on other chassis, including those of the FWD Corp., Spartan, and Simon Duplex. The move to other chassis builders allowed Baker to continue to provide its devices to fire departments.

In 1996, the company changed its name to Aerialscope Inc. It now produces and

delivers 75- and 95-foot (23 and 29m) aerials, and provides rebuilt and remanufactured units as well.

The company's current units are placed on either FWD or Spartan Motors chassis with cabs constructed of aluminum, stainless steel, or galvaneal steel. Platform capacity is rated at 1,000 pounds (454kg) for the 75-foot (23m) version and 800 pounds (363kg) for the 95-foot (29m) aerial.

Many big-city fire departments have come to rely on the aerial almost as much as their pumper and tanker units, and their dependence on aerials reflects the ever-changing nature of late-twentieth-century firefighting techniques.

The Miami-Dade Fire Rescue Department is a perfect example of a complex fire/rescue operation. Miami-Dade originated as a "fire patrol" service in 1935, then became a division of the Dade County Public Safety Department in 1958. But as early as 1965, a population explosion in the region prompted a reorganization, which

Above: *The Ogden City (Utah) Fire Department equips itself with a fully loaded 1988 Pierce-Dash with a 105-foot (32m) aerial. Aerials can be mounted midship or at the rear.* **Inset:** *A bird's-eye view of a hook-and-ladder demonstration during an exercise.*

resulted in the creation of the Metropolitan Dade County Fire Department. By 1973, a medical rescue program was started.

In 1996, the department responded to more than 154,000 calls, 70 percent of which were for medical rescues. Fire calls jumped by 10.7 percent from 1986 to 1996, with the fifteen-hundred-person department responding to 1,930 structure or building fires. Reflecting the continuing growth of the area, Metro-Dade now uses five aerial ladders, a 100-foot (30.5m) ladder truck, and two platform trucks.

Rising to such challenges along with E-One and KME is Pierce Manufacturing Inc. In early 1997, Pierce upgraded its aerial platform by enlarging platform space by 60 percent to a 22-square-foot (2 sq m) basket that accommodates four firefighters wearing full protective gear and carrying equipment and tools.

The platform features controls attached by a coiled cord, which can be held by the operator and placed in one of three fixed mounts on the basket. Pierce's LifeLadder is a new feature–a mounting bracket on the basket that allows a roof ladder up to 20 feet (6m) long.

Such improvements prompted the city of Richmond, Virginia, to sign a $12.6 million contract for thirty-one new Pierce trucks. The contract called for nineteen Quint combination trucks, each constructed on Pierce's Dash chassis and featuring a Hale pump with a 2,000 gpm (7,570lpm) rating and powered by a 470-horsepower Detroit Diesel Series 60 engine.

The Pocatello (Idaho) Fire Department chose to equip itself with this customized 1990 Pierce 1,500-gallon (5,678l), 500-gpm (1,893lpm) pump fire engine with a modest 55-foot (16.8m) aerial. Note the unusual paint scheme.

Fourteen of the nineteen Quints were equipped with 75-foot (23m) aerials and 500-gallon (1,892.5l) water tanks. Four were equipped with 105-foot (32m) aerials and 300-gallon (1,135.5l) water tanks. The last featured a 100-foot (30.5m) aerial platform with the new 22-square-foot (2 sq m) basket and 200-gallon (757l) tank.

To beef up its line in late 1997, Pierce purchased Nova Quintech, which manufactures the Sky-Arm, a four-section, 100-foot (30.5m) aerial ladder with articulating platform. Pierce's acquisition will also allow the company to offer the Sky-Five, a 102-foot (31m), five-section aerial that is mounted either midship or at the rear.

Most towns and small and medium-size cities today have ordinances that ban the construction of buildings higher than their departments' aerials can climb. But with the advent of 100-foot (30.5m) aerials, cities can build sizable office buildings, hotels, and government and medical facilities. By expanding office and hotel space in particular, cities can generate more sales and bed tax revenue.

While firefighting technology can't take all the credit for these financial boons for municipalities, it certainly helped pave the way.

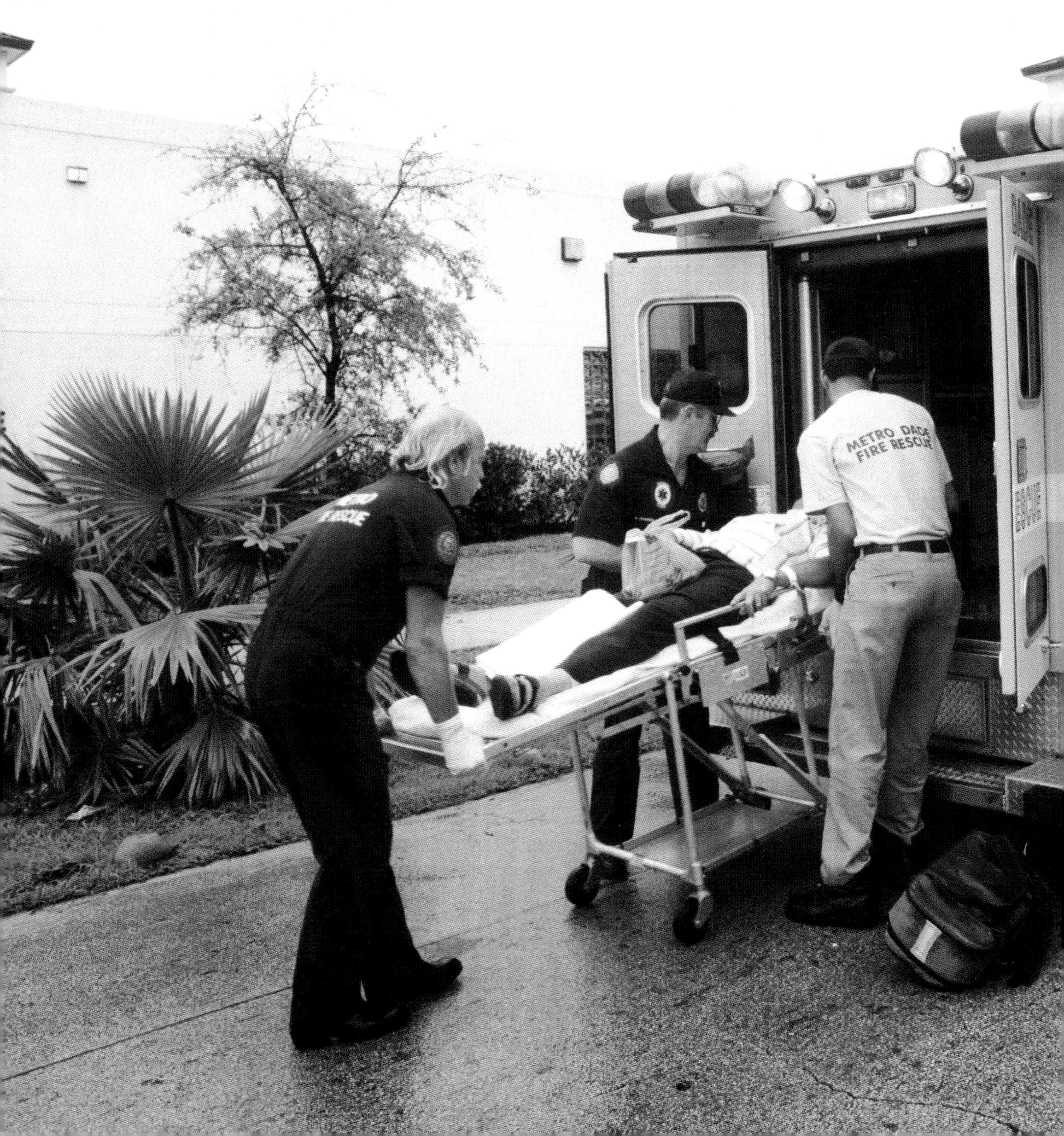
METRO DADE
FIRE RESCUE

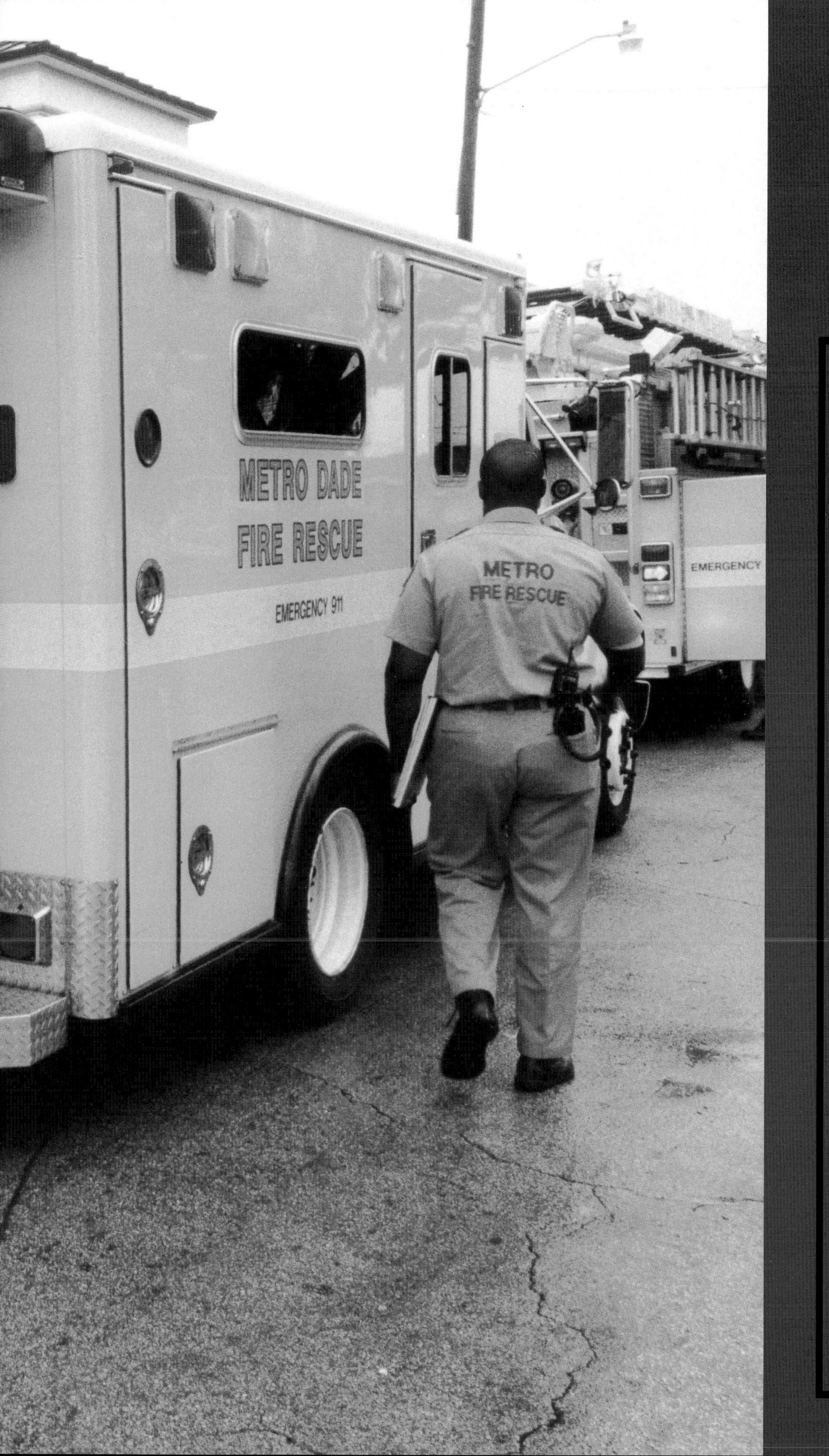

3

Ambulances and Rescue Units

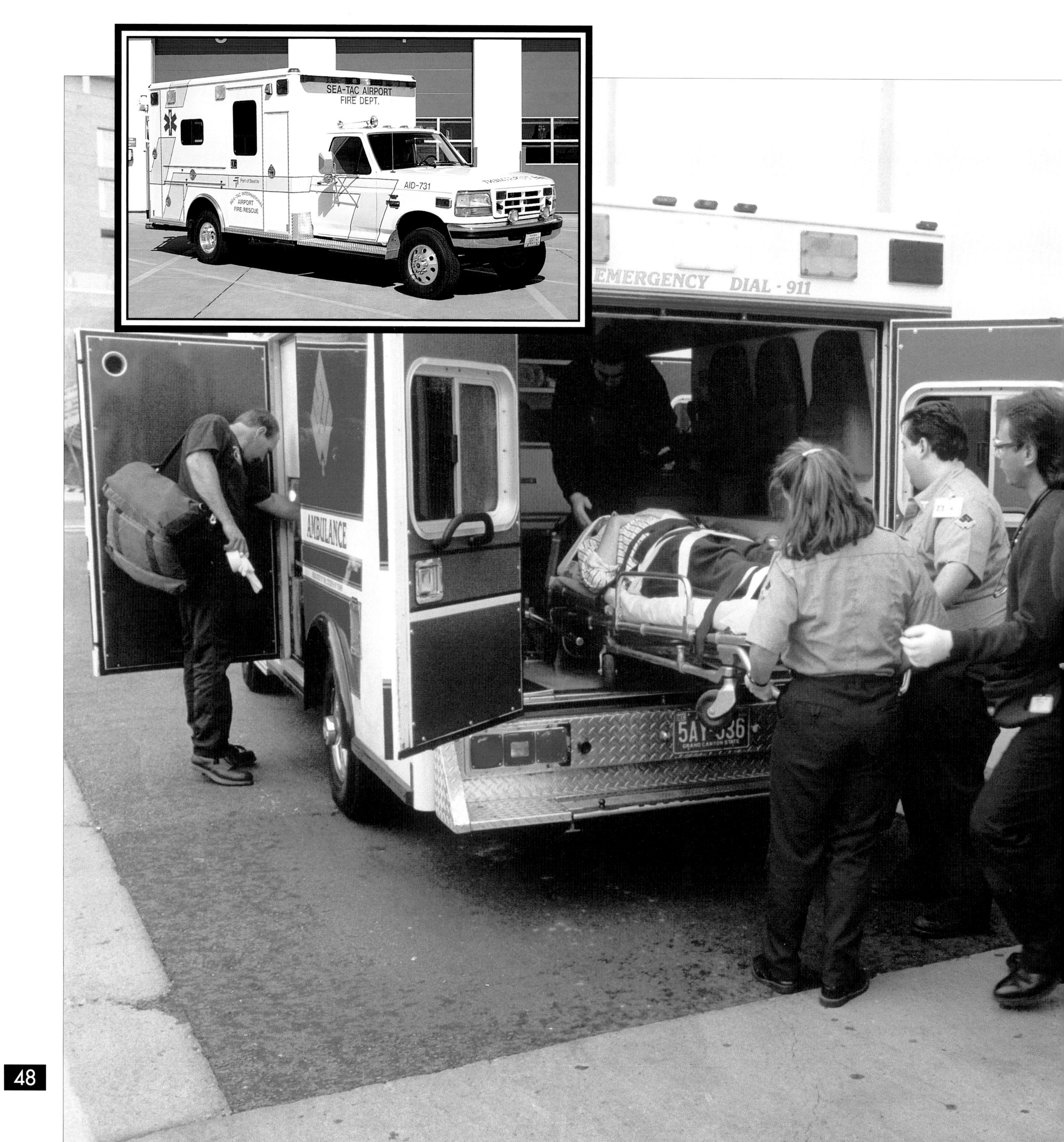
SEA-TAC AIRPORT
FIRE DEPT.
AID-731
SEA-TAC INTERNATIONAL
AIRPORT
FIRE RESCUE
EMERGENCY DIAL - 911
AMBULANCE

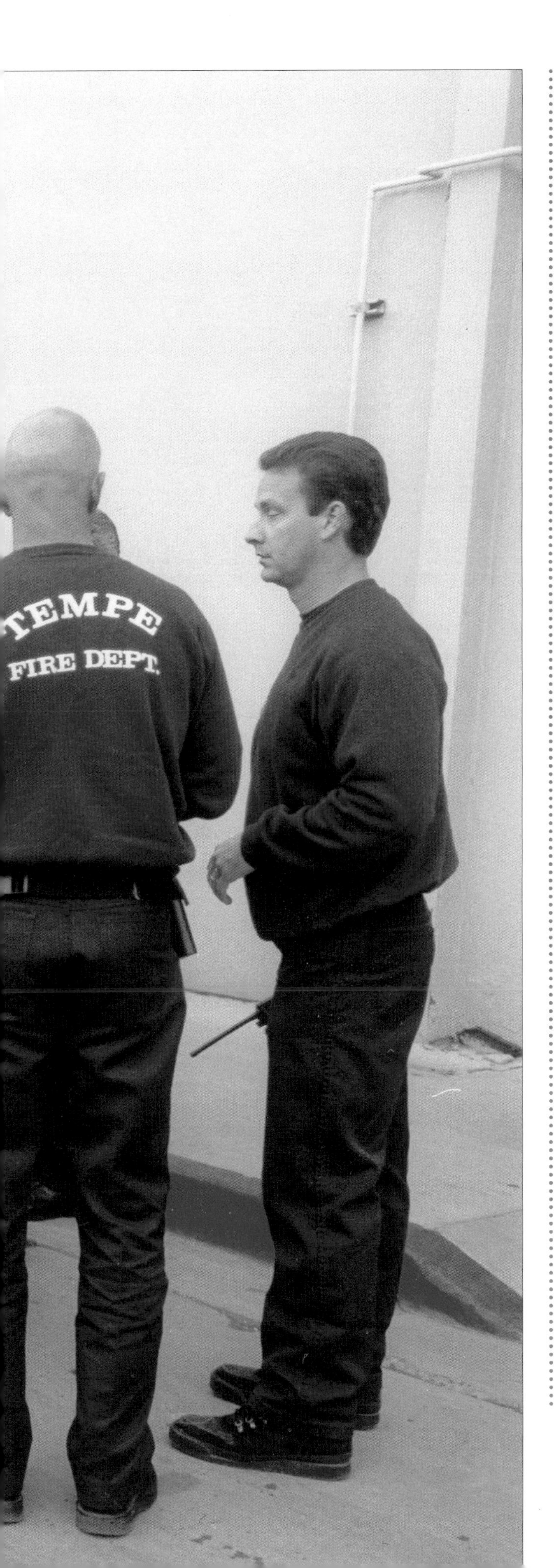

Previous pages: *Fire rescue units require a full staff to respond to any emergency.* **Left:** *Firefighting and rescue crews have removing victims from dangerous scenes down to an exact science.* **Inset:** *The Sea-Tac Airport Fire Department in Washington uses a 1996 F-Series 4×4 medic coach for crash duty.* **Above:** *The United States Marine Corps ordered a fleet of 1990 Pierce-Dash 4×4 rescue trucks. This unit was assigned to a Marine Corps base in Yuma, Arizona.*

▲ ▲ ▲ ▲ ▲ ▲ ▲ ▲ ▲ ▲

Nearly forty years ago, hospital physicians realized that many lives were lost due to accidents and illnesses because victims did not receive treatment quickly.

It was common practice for victims to be rushed to the hospital in a station wagon or even a hearse borrowed from the local mortuary.

The ambulance had long been a staple of emergency care in both North America and Europe. Two world wars had taught local jurisdictions that quick, specialized transportation to medical facilities saved lives. Doctors recognized, however, that precious time was lost between the time of an accident or the onset of an illness and the time the patient was delivered to the hospital emergency room. This was time that could literally mean life or death for the patient.

Through the 1950s, fire departments large and small were still preoccupied with structure fires and brushfires. As many as nine out of ten calls to community fire departments were for these fires. There were few rescues and extrications, no hazmat calls, and only rare emergency medical service. In the 1960s that changed–first due to the medical-evacuation technology gained during the Vietnam War, and second, in response to the civil unrest being experienced throughout the country. Fire departments were forced to confront a new age in public service: they had to save lives

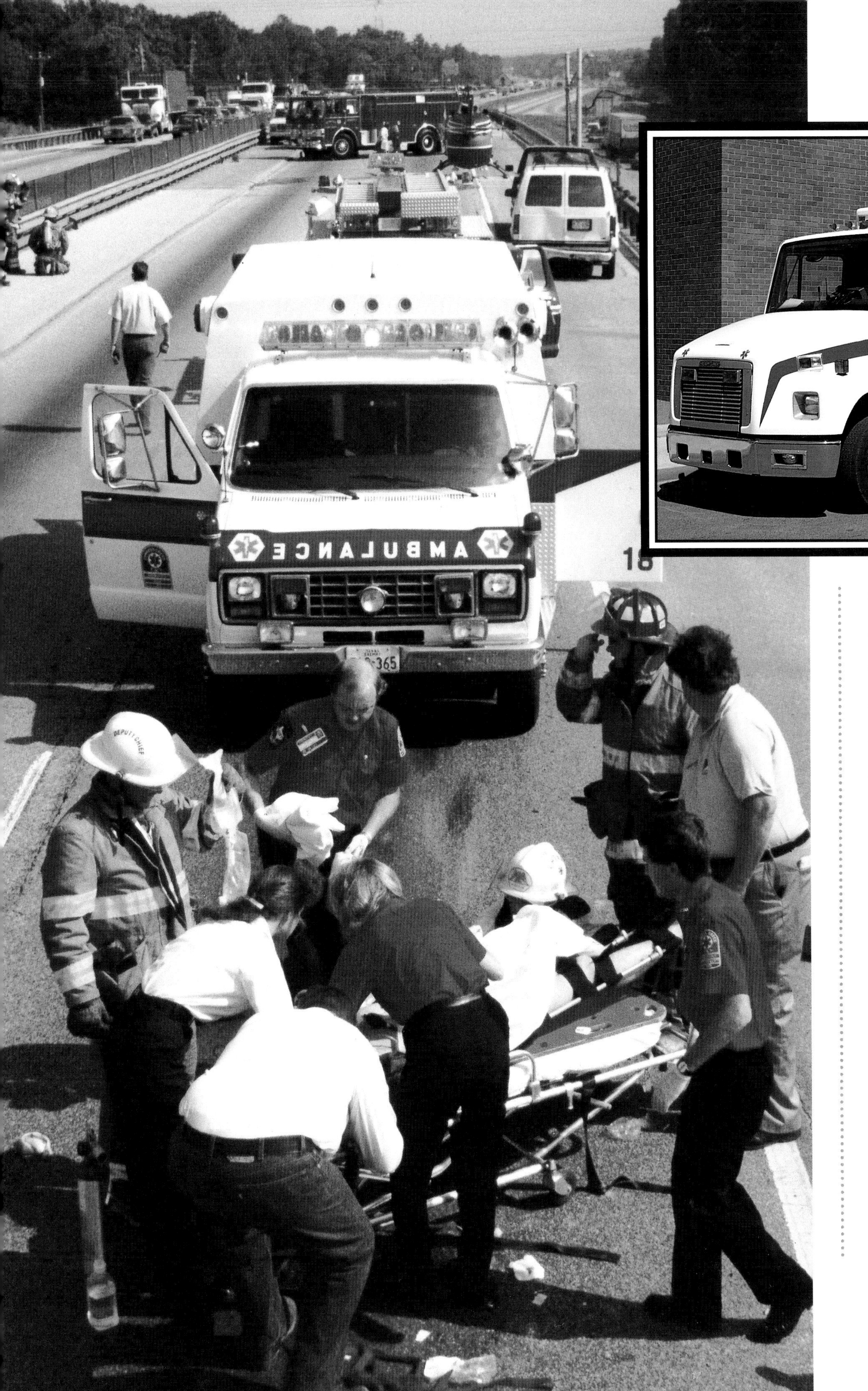

Left: *The survival rate for accident victims has increased ten-fold over the last two decades as firefighter and paramedic teams employ more sophisticated means to provide medical aid in the field.* **Above:** *The Bend (Oregon) Fire Department employs a 1993 Freightliner Lifeline medic rescue ambulance.*

▲ ▲ ▲ ▲ ▲

with early response. This became even more important as the need to battle structure fires dwindled.

The Independence (Missouri) Fire Department mirrors similar-size departments throughout the country. In 1996, it responded to 7,310 emergency medical calls. That's 64.8 percent of its 11,288 total runs. Structure fires accounted for only 9.4 percent of the department's emergency responses. The California desert community of Adelanto has even more startling figures: Adelanto firefighters responded to 1,273 medical calls–71.2 percent of its total run–in 1996. Only 2.5 percent of its calls for 1996 were for structure fires.

New Responsibilities, New Equipment

Demands for emergency medical aid require a more sophisticated operation and state-of-the-art equipment. It's no longer enough to transport victims from an accident scene to the hospital in less than five minutes. Rather, the hospital must come to the victim. In fact, the finest hospital emergency rooms of the 1970s couldn't come close to the technology that has become standard in today's fire rescue units.

Emergency One has developed a series of heavy-duty rescue units, rescue pumpers, and ambulances that boast top-quality construction and a fully equipped medical unit inside a custom interior.

These rigs are mounted on either a Navistar 4700LPX, Navistar 4700LP, or Freightliner Business Class chassis with a wheelbase ranging from 165 to 170 inches (419 to 432cm). Gross vehicle weight for the units–dubbed the Street Warrior Series, and the Trident Series for the ambulance rescue pumpers–ranges from 20,000 to 27,000 pounds (9,080 to 12,258kg). Engines offered are Navistar, Cummins, and Caterpillar diesels with horsepower ranging from 190 to 300. Overall length for these vehicles ranges from 273 to 285 inches (693.5 to 724cm).

The interior package for the Trident ambulance rescue pumper includes a full life-support system with an oxygen tank securing bracket that can be accessed under the squad bench through the right rear exterior compartment. It also provides a pair of Puritan Bennett oxygen outlets and a suction system with a high-volume suction pumper, regulator, and collector.

Its double-door construction on equipment compartments allows for safer access in highway or limited-space conditions. In all there is 140 cubic feet (4 cu m) of storage capacity, plus a transverse compartment with left- and right-side access for multiple spine board, stair chair, and scoop stretcher storage.

A 1990 Pierce-Dash, non-walk-in body rescue unit plies the roadways of Houston, Texas.

The Trident's fire suppression system features a 150-gallon (568l) polypropylene water tank, a 250 gpm (946lpm) pump, and a stainless steel pump panel behind a roll-up door.

E-One's Medilink mobile critical care unit for cardiac and neonatal care can accommodate two patients. Its equipment is tailored to hospital needs with a 110-volt/220-volt generator, air conditioning, heat, Grade E compressed air, liquid oxygen, a 1,200-pound (545kg) rear lift, hot and cold running water, a refrigerator, a microwave, a warming cabinet, an X-ray view box, and even a CD player, a television, and a VCR.

E-One's rescue units provide similar features and are constructed principally from aluminum. The Type I is on the T100 Ford F 350 XLT chassis or the T100 Chevrolet C3500 chassis. Using a van conversion on a truck chassis, the Type III models are placed on a T300 Ford E 350 XL chassis with a 138-inch (350.5cm) wheelbase or a T302 Ford E 350 XL chassis with a 158-inch (401cm) wheelbase.

Other rescue units from E-One's American Eagle services include the walk-in and non-walk-in rescue units. These trucks are placed on a Ford F 800 chassis with a 190-inch (482.5cm) wheelbase. They are powered by 250-horsepower Ford diesel engines through Allison MD3060P automatic transmissions. A major difference between the two models, besides walk-in capability, is storage space. The walk-in unit provides 330 cubic feet (9 cu m) of lighted and vented compartment space, while the non-walk-in unit has 670 cubic feet (19 cu m).

Rescue units can also be constructed with custom Cyclone II, Hurricane Tilt,

Opposite: *The fire department in Sunnyvale, California, is equipped with a 1993 HME-Quality 1500 rescue unit.* **Left:** *Firefighters inspect a diesel locomotive that derailed in inclement weather.* **Above:** *The LaPine (Oregon) Fire District uses a 1994 Freightliner with a Boise Mobile-Equipped body and a Marmon Herrington 4x4 system. It is also equipped with a 50-gallon (189l) foam tank.*

▲ ▲ ▲ ▲ ▲

The Jaws of Life hydraulic tool is an integral part of a firefighter's equipment box when removing victims from crushed vehicles. **Above:** *The Mid-Columbia (Oregon) Fire and Rescue Department remains loyal to local manufacturers by equipping its firefighters and rescue personnel with a 1990 Kenworth, built in the neighboring state of Washington. It offers a 750 gpm pump and 3,000-gallon (3785l) tank.*

▲ ▲ ▲ ▲ ▲

and rear-engine Hush chassis on commercial Navistar or Ford chassis.

KME recently produced a unique pair of high-tech custom rescue units with commercial chassis. The company delivered to Slatington, Pennsylvania, an International 4900 two-door cab and chassis unit with a 14-foot-6-inch (4.5m) walk-in unit constructed of 3/16-inch (4.8mm) aluminum body. It also features a 25,000-watt generator and a complete Holmatro rescue tool complement.

The Clarendon Hills (Illinois) Fire Department owns this 1994 Pierce-Dash rescue unit with walk-in body.

Delivered to the Vineland (New Jersey) Fire Department was the KME MFD Custom Contour Tilt Cab with an 18-foot-6-inch (5.5m) walk-around unit complete with roof compartments within its 3/16-inch (4.8mm) aluminum body. It also is equipped with a 25,000-watt generator and features a Will-Burt 60,000-watt light tower.

Amtech produces rescue and fire units that can be found on Spartan or International chassis, with walk-in rescue units ranging from 18 to 20 feet (5.5 to 6m) with compartments as deep as 28 feet (8.5m). The International models also feature electric reels with 200 feet (61m) of electrical cord.

The Type II TraumaHawk ambulance produced by American Emergency Vehicles is placed on a Ford E 350 2x4 chassis with a 138-inch (350.5cm) wheelbase and is powered by a 7.3-liter Ford Turbo Diesel V8 engine to generate 210 horsepower at 3000 rpm. The Type I TraumaHawk is essentially the same but is placed on a 161-inch (409cm) wheelbase.

Units manufactured by E-One, KME, and other makers give firefighters and medical crews all the tools necessary at the scene to not only fight fires but to provide virtually complete medical attention for victims. These units also serve as a command base for major incidents, minimizing the need for complex radio communications to faraway command centers.

13
RESCUE

4

Crash Trucks

Previous pages: *Crash trucks at civilian and military airports are the monsters of firefighting equipment. With a price tag of about a quarter million dollars each, these crash rigs are essentially pumper/tankers with all the bells and whistles and inflated to about three times the size of the standard fire engine. Their job is to douse a structure fire that is not an office building or residence but a cigar-shaped structure with hundreds of people inside.* **Right:** *Rescue crews conduct mop-up operations at an airport and aid the FAA in maintaining the integrity of the crash scene.* **Inset:** *In Seattle, Washington, a 1991 Oshkosh T3000 with a 3000-gallon (11,344l) tank and 410-gallon (1,552l) foam tank awaits action.*

▲ ▲ ▲ ▲ ▲

A rugged 1996 Emergency One Titan, the granddaddy of crash trucks. This one is equipped with a 1750 gpm pump, 3000-gallon (11,355l) water tank and 500-gallon (1,893l) foam tank. It can go from 0 to 50 miles per hour (80.5km) in forty seconds and cover terrain under any weather conditions.

Crammed onto a mere 7,800 acres (3,120 ha) of land are six runways that accomodate sixty-six million passengers each year at Chicago's O'Hare International Airport.

O'Hare is less than half the size of the Dallas/Ft. Worth airport and only one-quarter the size of Denver International Airport, yet it earns the distinction as the world's busiest.

The three crash stations at O'Hare are equipped with a dizzying array of specialized equipment probably not seen at any other international airport of comparable size and activity. In addition to the city's fire department, the U.S. Air Force also maintains a full complement of equipment ready to respond to any emergency.

O'Hare is no stranger to tragedy. Many of its current firefighters are veterans of one of the most horrific air crashes in U.S. history. On Friday, May 25, 1979, the beginning of Memorial Day weekend, American Airlines Flight 191, en route from O'Hare to Los Angeles, crashed. The DC-10, carrying 258 passengers and a crew of ten, took off shortly after 3 P.M. As it ascended, its number one engine pylon failed and the plane lost its engine. It slammed into the ground not far from the airport, killing all on board. The Chicago Fire Department staffing O'Hare was on the scene in moments.

Though its equipment was a product of late-1970s technology, it was not much different from what we would expect

Above: *The U.S. Air Force puts firefighters through their paces in a training exercise at an Air Force airport facility.* **Inset:** *A 1-ton (.907t) Dodge dual cab truck serves as a rescue unit.*

based on today's standards. Chicago's fire department has always equipped O'Hare with the very latest in fire suppression technology.

Today, O'Hare is equipped today with a pair of 1992 Emergency One pumpers with 50-foot (15m) Squirts, a 1988 E-One 95-foot (29m) ladder truck, a 1988 Spartan Super Vac squad, a command van, three ambulances, an aircraft stairway, and nine crash rigs. Big, brawny, and with a price tag of about $250,000 each, these crash rigs perform the lion's share of the duties at major disaster scenes.

Most crash rigs operated by the Chicago Fire Department today are Oshkosh T-3000s equipped with 3,185 gallons (12,055l) of water and 420 gallons (1,590l) of foam. One Oshkosh T-3000 is equipped with a Snozzle, an appliance mounted at the end of a boom to blast a stream of water or foam right through the thin structure of an aircraft, allowing it to knock down the blaze inside.

The U.S. Air Force also keeps an impressive line of equipment. A pair of Quality P-2 crash rigs with 2,000 gallons (7,570l) of water and 205 gallons (776l) of foam stand at the ready. Also on standby are a Ford F-350 Rapid Intervention vehicle that carries foam, a Chevrolet C-30 rescue vehicle, and two Oshkosh rigs. One is a P-4 crash rig with 1,500 gallons (5,677.5l) of water and 105 gallons (397.5l) of foam, the other a P-19 crash rig with 1,000 gallons (3,785l) of water and 130 gallons (492l) of foam.

The Mesa (Arizona) Airport Fire Department employs an aged, but efficient, 1965 American LaFrance with 750 gpm (2,839 lpm) pump, 1150-gallon (4,352l) water tank and 150-gallon (568l) foam tank.

Airport Emergencies

Consider the dynamics of fighting an aircraft fire and the prospect of rescue under the most arduous conditions. A Boeing 727 has a wingspan of 108 feet (33m) and measures 153 feet (46.5m) in length. It's powered by three Pratt & Whitney turbofan engines. Inside are not only 189 passengers and a crew of about ten, but 9,800 gallons (37,093l) of highly volatile aviation fuel. Add inclement weather, like a snowstorm or triple-digit heat, and the firefighter's job borders on herculean.

A crash rig is essentially a pumper/tanker with all the bells and whistles, inflated to about three times the size of a standard fire engine. Its job is to douse a structure fire that is not a building or residence, but a cigar-shaped structure with hundreds of people inside.

These crash rigs can rush to the scene of an airplane crash over uneven terrain in mud, slush, or snow at a top speed of 65 miles per hour (104.5km). They can fight a blaze without a single firefighter even leaving the vehicle. Fighting a fire by remote control from inside a crash tender allows the firefighters to assess damage, casualties, and danger before exposing themselves to the fire.

Above: *A 1988 Oshkosh P-15 with a pair of 1250-gallon (4,731l) water tanks, a 6100 gpm (23,089l) pump, and a 515-gallon (1,949l) foam tank serves the McChord Air Force Base in Washington state.* **Opposite:** *A crash truck performs duty at the Youngstown Air Force Reserve Base in Ohio.*

Equipment Inventories

The inventory at the Dallas/Ft. Worth International Airport Department of Public Safety is not much different from O'Hare's. At Dallas/Ft. Worth, the fire department is equipped with twelve crash trucks–eight for front-line duty and four in reserve. All are E-One HPR 300s equipped with a 3,000-gallon (11,355l) water tank, foam capacity of 385 gallons (1,457l), a roof and bumper turret, and a pair of hand lines, according to Division Commander Alan Black. Black was at two air disasters at Dallas/Ft. Worth in 1985 and 1988, in which crash trucks played the lead role in fire suppression and rescue.

On August 2, 1985, Delta Flight 191, carrying 156 passengers and eleven crew members, crashed in a thunderstorm after the Lockheed L-1011 clipped a car on the highway at the edge of the airport. Killed in the crash were 137 passengers, the crew, and the driver of the car. On August 31, 1988, Delta Flight 1141–a Boeing 727–stalled on takeoff and crashed, coming to rest about 3/4 mile (1.2km) from the end of the runway. Fourteen people were killed.

At that time, the fire department was equipped with Oshkosh T-1500s and

M-15s. Firefighters responded to the wrecks in less than a minute. The crash tenders quickly knocked down fires at both accidents and performed superbly, Black said. He noted that the performance of the mid-1980s Oshkosh crash trucks and of the current crop of E-Ones is almost identical.

Across town is Love Field, where the Dallas Fire Department operates three front-line crash trucks, according to Fire Captain George Freeman. The department operates a 1981 Oshkosh 3000 that has a 3,000-gallon (11,355l) water tank and capacity for 310 gallons (1,173l) of foam. Its 1986 Oshkosh 1000 has a 1,000-gallon (3,785l) water tank with a 130-gallon (492l) foam capacity. A much older model remains in operation for front-line work, Freeman said. The department's 1972 Kline holds 7,000 gallons (26,495l).

Another facility with an extensive inventory of equipment is the Naval Air Station at Patuxent River, Maryland. Located 65 miles (104.5km) south of the Pentagon, the station overlooks Chesapeake Bay on 7,950 acres (3,180ha). It houses the U.S. Navy's only test pilot school and has been the site of aircraft testing, evaluation, and research since 1943.

The naval station's fire and emergency services protect more than sixteen thousand military and civilian personnel. Its 160 aircraft are spread over 1,064

Above: *A volatile tanker fire is doused with a heavy spray of foam. Airport fires often require joint responses from civilian and military rescue teams.* **Left:** *An 8×8 rig at the Youngstown Air Force Reserve Base in Ohio. Crash trucks similar to this hold more than 3,000 gallons (11,355l) of water.*

▲ ▲ ▲ ▲ ▲

buildings, ten double-bay hangars, and nearly 60 miles (96.5km) of road.

Fire protection at the naval station is a massive operation that rivals that of any medium-size city. Seventy-three uniformed firefighters man seventeen apparatus, with nearly 70 percent of the station's calls to the airfield. Among the equipment used daily are half a dozen Oshkosh crash tenders. Three are P-19 models that carry 1,000-gallon (3,785l) water tanks; the other three, purchased in 1992, are TA-3000s that carry 3,000-gallon (11,355l) water tanks.

By contrast, the Detroit Metro Airport Fire Department equips its firefighters with three Oshkosh machines. It has a pair of 1982 Oshkosh M-15 crash trucks and a 1988 DA-1500 dual-agent crash truck. Its line of pumpers includes a 1994 Pierce Class A pumper and the 1987 GMC/Pierce mini-pumper.

The granddaddy of aircraft rescue vehicles is perhaps the Titan HPR 8x8, such as the one assigned to Denver International Airport. Sitting on a frame constructed of low-carbon tubular steel rails and with a 212-inch (538.5cm) wheelbase, the rig features a 3,170-gallon (11,998l) water tank and a 504-gallon (1,907.5l) foam tank. An optional water tank can be ordered, measuring 2,642 gallons (10,000l) and with a foam tank of 337 gallons (1,275.5l).

The engine is a rear-mounted 585-horsepower Detroit Diesel 8V92TA with a 1,500-square-inch (9,678 sq cm) radiator that would dwarf that of any eighteen-wheeler tractor trailer rig. Power is transferred through an Allison HT-750DR five-speed automatic transmission. The cab is a two-door model mounted on six insulated rubber cushioned mounts. It has a 3,700-square-inch (23,872.5 sq cm) tinted windshield and roll-down tinted door windows. It has large enclosed compartments that allow storage for hose lines, reels, and ancillary firefighting equipment. Outside the station doors, the Titan 8x8 can hit a respectable speed, going from 0 to 50 miles per hour (80.5kph) in about forty seconds.

A 6×6 rig for military purposes at the Rickenbacker Air National Guard Base in Ohio.

A smaller Titan, but no less imposing, is the 1995 E-One Titan IV used by the England Airpark/Rural Metro Fire Department of Alexandria, Louisiana. The agency serves about three thousand people in a 6-square-mile (15.5 sq km) area and operates out of a single station. Its Titan 4x4 carries 1,057 gallons (4,000l) of water and 135 gallons (511l) of foam, and is powered by a 585-horsepower Detroit Diesel.

A slightly larger operation gives the Keesler Air Force Base Fire Department in Mississippi three crash tenders. The department protects more than twenty-two thousand people over 33 square miles (85.5 sq km) out of one station. Along with an airfield is a full regional medical center, plus nine housing units. It also has mutual aid agreements with nearby Dílberville, Biloxi, and Gulfport. Twin 1985 Oshkosh P-10 crash trucks are assigned to the airfield along with a 1995 Teledyne four-wheel-drive P-23 crash truck. These three units are among eleven apparatus operated by Keesler.

In Columbia, South Carolina, the Columbia Metropolitan Airport Crash

A 1993 Freightliner-Fire Bann equipped with a Boeing Renton with a 2000 gpm (7,570l) pump, 5000-gallon (18,925l) water tank and 615-gallon (2,328l) foam tank.

Fire Rescue Department has just a single KME pumper to protect two thousand people in a 6-square-mile (15.5 sq km) area, but it also has three crash trucks. The Columbia department's responsibility is to cover a heavy commercial area out of one station with an emphasis on emergency medical services, which accounted for about one-quarter of its 304 calls in 1995. The fire department also trains its crew in confined-space rescue, high-level rescue, haz-mat operations, and even radiological detection. Its three crash trucks–all carrying 1,500-gallon (5,677.5l) water tanks–are 1985 and 1986 Oshkosh T-1500s and a 1974 Walter BDQG-1500.

The Ronaldsway Airport (Isle of Man) in the United Kingdom provides an operation not much different from its American counterparts, with three heavily equipped crash trucks. This airport occasionally accommodates big aircraft, including Boeing 737s, 757s, and 727s. One officer and six firefighters operate the facilities out of a total Rescue and Fire Fighting (RFF) complement of four officers and eighteen firefighters.

The airport was built in the 1930s and began accommodating De Haviland aircraft. By 1935, it had experienced its first crash, involving a De Haviland pilot and six passengers. The De Haviland crashed on takeoff, but there were no fatalities. The accident sparked the Air Ministry to purchase a Morris van from a local confectioner and convert it to a fire tender.

The Royal Air Force took over Ronaldsway Airport operations during World War II and built runways. In 1948, the airport returned to civilian use. Twelve years later, Viscount aircraft began using the facility, signaling a need for a larger operation and a better-equipped fire department. In 1961, a three-bay fire station was constructed, allowing all fire vehicles at the airport to be housed together for the first time. By 1990, five bay stations were in operation.

One of the airport's heaviest pieces of equipment is the Carmichael Jet Ranger 200. It's constructed on a Shelvoke and Drewry four-wheel-drive chassis and powered by a 12-liter Detroit Diesel engine. It carries 2,140 gallons (8,100l) of water and 238 gallons (900l) of concentrated foam. The foam and water are dispensed through a C7000, a 889 gpm (3,364lpm) plumbing system.

The Carmichael Jet Ranger 200 is also equipped with massive 500-watt floodlights on the roof, a pushup ladder mounted on the roof, and two sets of breathing apparatus with spare cylinders. Inside its compartments are ropes, gloves, blankets, torches, and hand tools, including a saw, to help with extrication.

The fire department's Simon Access Protector is placed on a Simon 195-inch (5,000mm) four-wheel-drive chassis and also powered by a 12-liter Detroit Diesel. The Protector carries 1,612 gallons (6,100l) of water and 198 gallons (750l) of concentrated foam applied through a 846-gpm (3,200lpm) dual-output monitor. It carries four 500-watt floodlights atop its roof, as well as a 110-volt portable generator to power its hand tools.

The Tennessee Air National Guard operates this 1987 Oshkosh P-19 crash truck at the Memphis International Airport. The apparatus is equipped with a 1000 gpm (3785 lpm) pump and 1000-gallon (3,785l) water tank. It's also equipped with 130 gallons (492l) of foam and 500 pounds (187kg) of Halon extinguishing agent.

The agency's third crash truck is a Simon Access Defender built on a Mercedes-Benz 148-inch (3,800mm) wheelbase with four-wheel-drive. It's powered by a Mercedes-Benz OM.424.1 diesel engine. This vehicle carries 1,189 gallons (4,500l) of water and 143 gallons (540l) of concentrated foam, which are dispensed through a 608 gpm (2,300lpm) system. It's also equipped with 500-watt floodlights, a triple pushup ladder mounted on the roof, and a complement of tools identical to its two brothers'.

Ninety percent of a crash rig's emergency calls amount to standbys as distressed craft come in for a rocky landing. But then there are the occasions when the waiting rig will respond to a crash and make the difference in saving lives. Lives were saved at the Dallas/Ft. Worth airport because the crash rigs were on the scene within seconds, not only to knock down flames but to extricate passengers and crew quickly. Without these rigs, the death toll from such crashes would be incomprehensible.

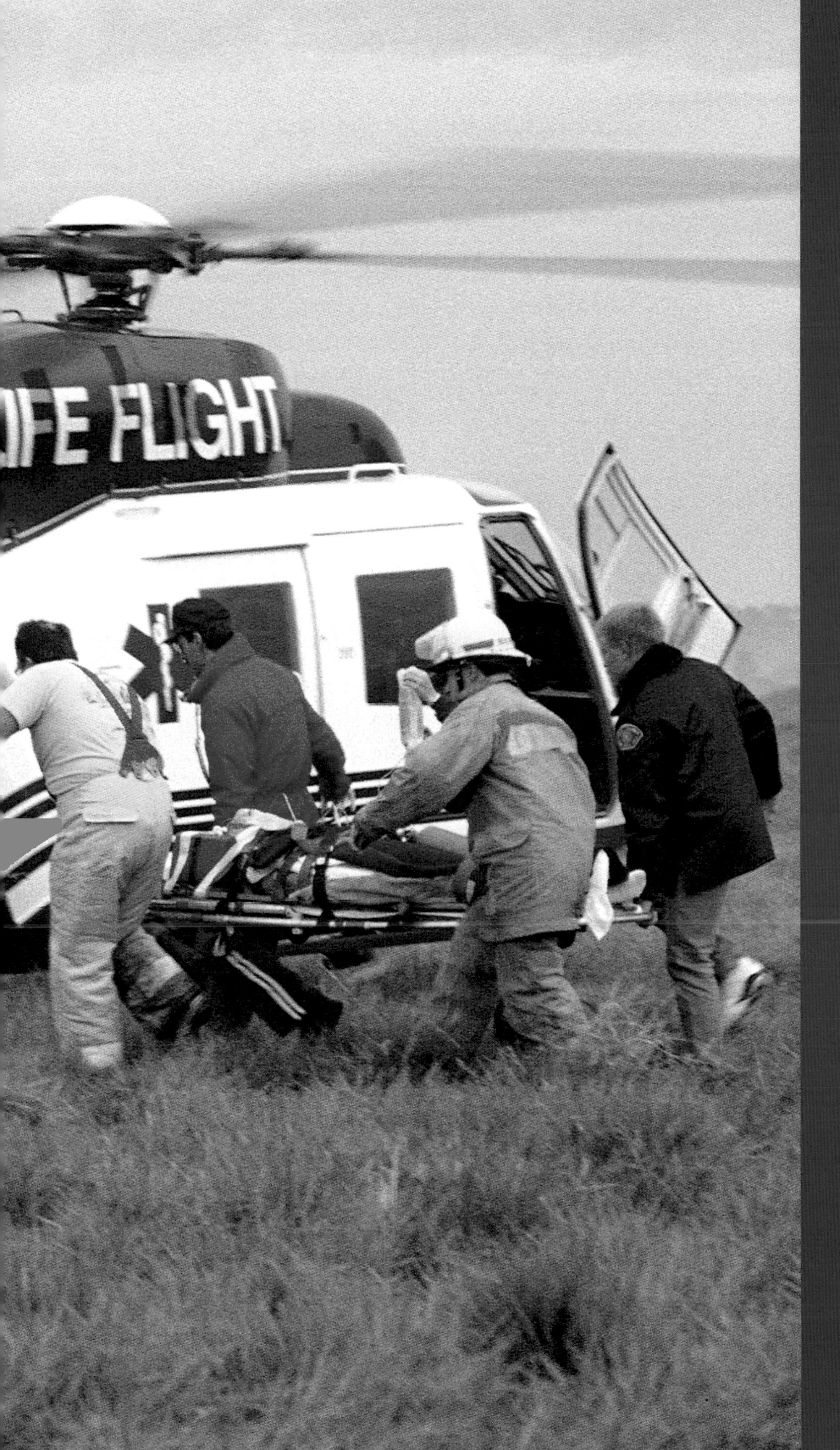

5

AIRCRAFT

Previous page: *As ground rescue services to remote areas become more sophisticated, so does air rescue, especially with helicopters poised to respond to rescues in difficult terrain.* **Right:** *An Eagle One search and rescue helicopter lifts off near several ground units during a rescue operation near Meriden, Connecticut.* **Opposite:** *A Jayhawk Sikorsky employed by the United States Coast Guard for service at Elizabeth City, North Carolina.*

▲ ▲ ▲ ▲

The radio traffic meant little to the members of the Marin County (California) Sheriff's Department Search and Rescue Team. They had arrived at Yosemite National Park on July 11, 1997, for their annual trip to the valley. They weren't there to perform rescue work.

To the team members, the radio chatter the day after their arrival was all too familiar. A climber had injured himself on El Capitan and the Yosemite Search and Rescue Team (YOSAR) was preparing for the flight up for the rescue. But by the time lunch was over, YOSAR had asked for assistance.

The Marin County team suited up in Nomex flight gear and rescue harnesses and hopped aboard a UH-60 Blackhawk helicopter for the short trip to the summit of El Capitan. When they arrived a few minutes later, they were greeted by YOSAR members. After a brief assessment of the situation, the team members descended 650 feet (198m) to a small ledge.

The team quickly surmised that the climber was unable to reach the summit on his own, so he was brought up on a litter. It was dusk and the team had to spend the night with their patient before the Blackhawk returned in the morning. The Blackhawk transported the patient and his rescuers to the valley floor, where another chopper was waiting to transport the climber to a nearby hospital.

The rescue was routine from start to finish. No one broke a sweat. The use of the Blackhawk typified the "work mule" duty that choppers now perform, not only in rescue operations but in fire suppression and in moving rescue teams and firefighters to inaccessible areas.

Choppers and fixed-winged aircraft have become necessary components of fire protection, rescue, and law enforcement work in both urban and rural areas. The Los Angeles County Fire Department was an early pioneer in air support, starting its air operations in 1957, when it purchased a Bell 47 helicopter from the Los Angeles County Sheriff's Department. The Los Angeles (city) Fire Department began its air operations unit in 1962, also by acquiring a Bell 47. And the San Bernardino County Sheriff's Department, which is responsible for law enforcement and mutual aid in the largest county in the United States, began its aviation division in 1971 with a pair of Bell 47s.

Novelties in the 1950s and 1960s, choppers now are crucial tools for dampening blazes, tracking criminals, and rescuing stranded hikers or motorists from rain-swollen creeks.

The possible usefulness of the chopper was first realized during the Korean War, when the egg-beaters performed exceptionally well in transporting wounded combat soldiers from the battlefield to M*A*S*H units situated just a few miles from the front lines. Civilian ambulatory services in those early years were in their infancy, but fire departments and law enforcement agencies saw the need for observation and command at incidents.

A decade later, the performance of the Huey cemented the chopper's position as a rescue unit as it saved the lives of thousands of soldiers in Vietnam. Many early law enforcement chopper pilots were Vietnam veterans. It should also be no surprise that most of the choppers that now see service in fire and police agencies are Army surplus.

Blackhawk and Bell

UH-60 Blackhawks similar to the one used by YOSAR and the Marin County Sheriff's Department to rescue the injured climber are the U.S. Army's utility helicopters. Used primarily as a combat troop assault vehicle, it can carry eleven to twenty-two

combat-equipped troops into battle. It can also move equipment and sling-load light vehicles to the battlefield. Configured as an air ambulance, it can carry up to six litter patients or seven ambulatory patients. It can be specially configured as a medevac helicopter with modern medical interior and avionics. The Blackhawk first saw combat when U.S. troops invaded Grenada in October 1983. It subsequently served in Panama, Haiti, Bosnia, and Somalia.

It accommodates a crew of four and–depending on its equipment and configuration–can reach speeds of up to 142 knots and an overall range of 359 miles (578km). The current production UH-60L is powered by a pair of General Electric T700-GE-701C turboshaft engines generating 1,700 shaft horsepower. It has a four-bladed main rotor and four-bladed tail rotor. It sits on a tricycle landing gear (one aft and two fore).

Bell helicopters have been used by many agencies since the 1950s, and are considered the most reliable and efficient choppers in the air. The Los Angeles County Fire Department upgraded in 1967 from the old Bell 47 to a Bell 204-B, which was equipped with a 320-gallon (1,211l) drop tank that allowed the pilot to deliver more water or retardant per hour to a fire scene than the conventional fixed-wing air tankers. The success of the Bell 204-B led to the department's acquisition of a full fleet of choppers. The department now operates a single Bell 206-B III Jet Ranger, three Bell 205A-1s, and four Bell 412s.

Opposite: *This helicopter is equipped with a water rescue basket for plucking stranded victims out of the water. This chopper is used by the Washington, D.C., Metro Police Department.* **Above:** *A Bell Jet Ranger chopper. These Jet Rangers are popular for police work. A Bell Jet Ranger can serve as an air ambulance with a cruising speed of 100 knots and light gross weight at 3,200 pounds (1,194kg). It seats five people and carries two litters.*

The department's Bell 206-B fills the need for an air ambulance thanks to its cruising speed of 100 knots and light gross weight of 3,200 pounds (1,453kg). It seats five people and carries two litters. The Bell 205A-1 was sought by Los Angeles County authorities because of its large seating capacity, internal litter-carrying capability of up to three patients, hoist-rescue capability, and performance at high altitudes on hot southern California days.

In all, the 205A-1 can carry up to fifteen personnel. Its turbine-powered engines

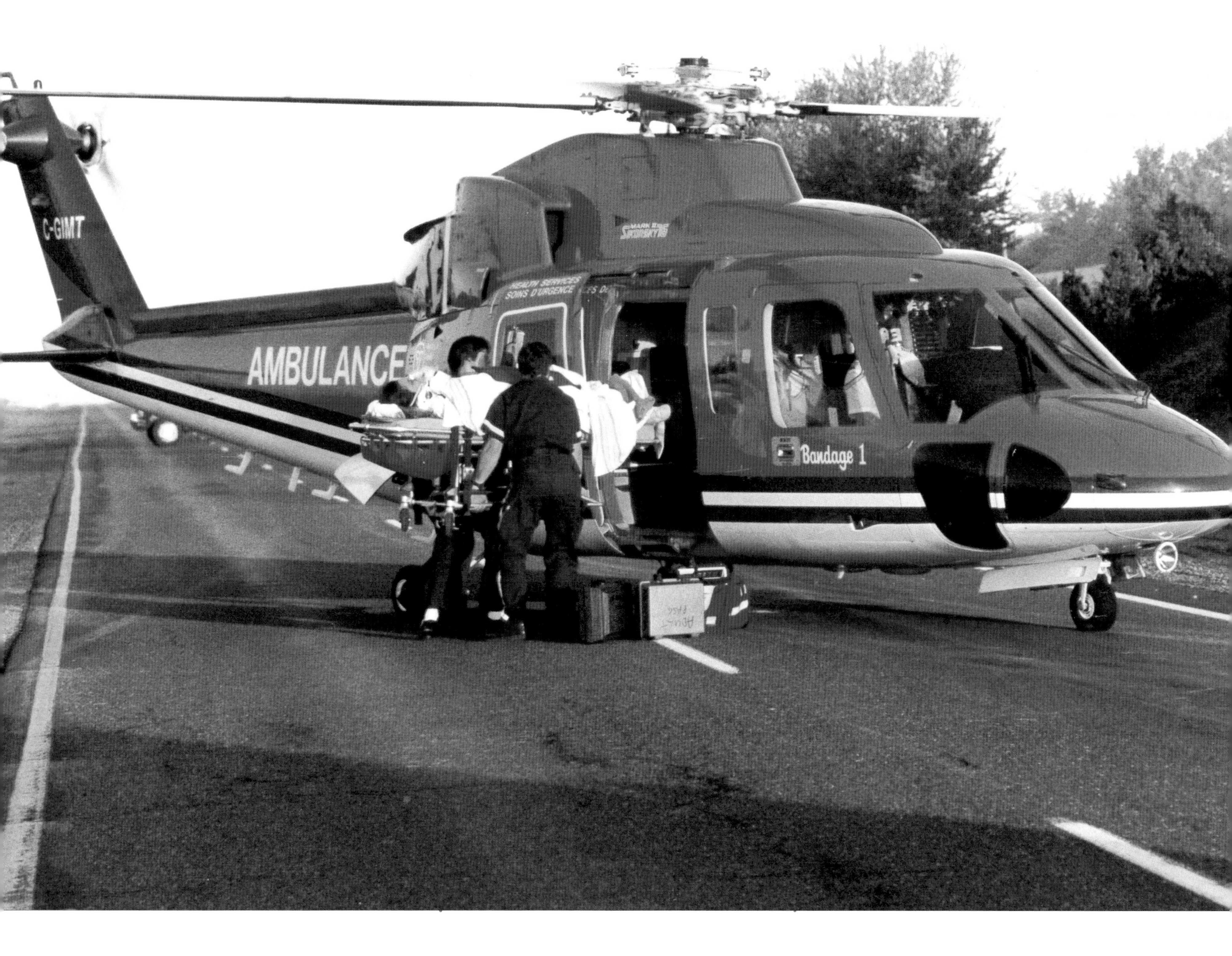
C-GIMT
AMBULANCE
HEALTH SERVICES
SOINS D'URGENCE
Bandage 1

generate 1,400 shaft horsepower to achieve a cruising speed of about 100 knots. Its gross weight is 10,500 pounds (4,767kg).

The 205A-1 will eventually be phased out to make way for the next-generation chopper, the Bell 412. The Bell 412 is perhaps the most sought-after helicopter by law enforcement and fire agencies because of its larger size, greater capacity to carry water and fire retardant, and overall superior performance.

Choppers in Action

The Queensland (Australia) Emergency Services team owns three choppers and has swatted down scores of blazes using the Bell 412 in a variety of missions. Like the Bell 205A-1, it carries up to fifteen people and three litters. It has a water tank capacity of 360 gallons (1,362.5l), but unlike the 205A-1, it is powered by a pair of Pratt & Whitney PT6-3 turbine engines that generate 1,800 shaft horsepower for a cruising speed of 120 knots or 149 miles per hour (240 kph). It can reach heights up to about 20,000 feet (6,096m).

Standard equipment for the 412 is a three-person rescue hoist with a 29 1/2-foot (9m) cable, up to four full-access stretcher positions, dual oxygen and suction systems, a 30 million-candlepower Nightsun searchlight, and a complete emergency service and public utility communications system with mobile telephone. The 412 can also carry a 396-gallon (1,500l) firebombing bucket and spray chemicals to assist in the control of oil spills. The sophisticated design of the 412 allows emergency crew members to reconfigure the interior of the chopper to serve a variety of medical and firefighting functions.

In Brisbane, Australia, pilots took delivery of a Bell 412 in August 1997 to replace an aging twin-engine Squirrel helicopter that had been in operation since 1984. The 412 saw action almost immediately after its acquisition. The Queensland Emergency Services team was scheduled to give local firefighters a demonstration of the new machine when it was called to respond to a plane crash in which five parachutists were critically injured.

At the scene of the demonstration, the Queensland crew immediately reconfigured the interior of the plane to accommodate the injured and flew to the scene. Some of the injured parachutists were treated by medical personnel from the Lifeflight section of the Royal Brisbane Hospital; the most critically injured were transported to the facility. In the same shift, the rescue team had to reconfigure the interior of the 412 again to serve as a neonatal ambulance, interhospital transport, and the primary on-scene accident vehicle.

Perhaps the two most important pieces of equipment for pilots are the wire strike protection kits for after-dark work and the Forward Looking Infra-Red (FLIR) observation system. In one incident, the Bell 412 was called to a scene in which a car had run off a treacherous mountain road. Ground crews had searched for the vehicle for three hours without success. The Bell 412 was brought in and crew members searched the area with its FLIR system. They discovered the car only 65 1/2 feet (20m) from the road. The search could not have been made with conventional flashlights and searchlights, but FLIR's infra-red night vision made it possible.

About half of Queensland's missions are for air medical services, with about 80 percent of those calls dedicated to interhospital transfers and the remaining 20 percent dedicated to on-scene accident responses. A mission crew for the 412 consists of a pilot, an ambulance officer, and

Air ambulances today are state-of-the-art, with virtually full-service medical teams and on-scene medical equipment to provide instant aid to injured victims.

▲ ▲ ▲ ▲ ▲

This 30 million-candlepower Nightsun searchlight allows observers in choppers to find even the most brush-obscured car wreck or lost hiker in the blackest of nights.

a crewman. For interhospital transfers, the crew will pick up a doctor and nurse from one of the Brisbane-based hospitals.

For the Los Angeles County Fire Department, most Bell 412 services are in mountainous and rural areas of the county. The department still covers a significant portion of urban territory–about 2,165 square miles (5,607 sq km)–but the Bell 412 and its sister airships are rarely involved in structure fires unless brush areas are threatened.

Typically, Los Angeles County has its helicopters on standby with firefighting crews during daylight hours in the May-to-October fire brushfire season. The Bell 412, Bell 206-B III Jet Ranger, and Bell 205A-1 are employed to land firefighting crews at brushfire scenes. After dropping off the crews, the choppers rendezvous with an engine company at a helispot near the fire. The county has more than one hundred helispots at strategic locations in brush areas that allow the choppers unlimited water access from fire hydrants. The engine company provides the hose lay to pump the water to the helicopter. The 360-gallon (1,362.5l) tanks for the Bell 412 and Bell 205A-1 take about one minute to fill, with average turnaround on drops about five minutes. This allows the chopper to drop more than 4,000 gallons (15,140l) of water on a hot spot in less than an hour, exceeding the performance of fixed-wing air tankers.

Like the Queensland operation, Los Angeles County firefighters make extensive use of the FLIR system for mapping, aerial reconnaissance, and moving equipment and personnel to a fire scene. The department also installed a snorkel on one of its Bell 205A-1s for quicker attacks on blazes.

Los Angeles County's air ambulance service uses the Bell 205A-1s and the 412 air squads, and each is staffed by a pilot and two firefighter paramedics. Each pilot is required to have four thousand hours of helicopter pilot-in-command time. About fifteen hundred of those hours are dedicated to training in mountain operations above 4,000 feet (1,219m). One squad is staffed twenty-four hours a day, the other only during daylight hours. In 1985, the county developed its Reserve Physician Program, which allows trained emergency physi-

Small craft like this aid observers for the California Department of Forestry in Hemet Valley, California.

cians to work weekends and holidays to provide a high level of emergency medical care without increasing the department's operating costs.

The air teams respond to remote areas where motorcycles, hang gliders, all-terrain vehicles, four-wheel-drives, aircraft, boats, and tractors have been involved in accidents.

The Los Angeles (city) Fire Department uses a pair of Bell 205s, originally designed for troop transport and medevac during the Vietnam War, as water droppers, for grass and brushfire suppression. The department's two Bell 412s are used as air ambulances and for hoist work and water dropping.

Initial training for pilots for the department's air operations is two hundred hours of instruction in basic operations, mountain terrain, heavy-load operations, confined-space landing, and aircraft emergencies. About one-third of the pilot applicants are successful in this training. The department is also phasing in a transition from the Bell 205s to the 412s.

The Los Angeles helitac crew is stationed at Fire Station 90 near the Van

Nuys Airport in the San Fernando Valley. The crew consists of a captain, an engineer, an apparatus operator, and a firefighter. Typically, the LAFD will spend $1,637 to deploy one four-person helicopter crew for twelve hours. The Los Angeles County fire crew spent $10,000 in wages for a rescue team and thirty-six firefighters to rescue a horse stranded on a sandbar in the Tujunga Wash area of Los Angeles during a 1998 winter storm.

Desert Patrol and Rescue

In San Bernardino County, the sheriff's department provides law enforcement support, search and rescue, fire suppression, and transportation services throughout the county. The department operates a dozen choppers: one Boeing MD 600N, four Boeing MD 500Es, two Bell 206L-3s, two OH-6A military surplus vehicles, a Bell 212 twin Huey, and two Bell UH-1H military surplus Hueys.

During 1997, the department's staff of fourteen pilots and five observers—plus nearly one hundred volunteer pilots, medics, observers, crew chiefs, and others—responded to 177 rescues and transported more than two hundred critically injured

victims in 265 hours of flying time. The department's aviation division, based out of Rialto in the western portion of San Bernardino County, aided fire agencies in battling seventy-eight fires for a total of 155 flying hours for the 1997 fire season.

The Bell 212 and the UH-1H Hueys are primarily used for search and rescue, fire suppression, SWAT (Special Weapons and Tactics), disaster relief, and general transportation duties. The Bell 212 carries up to thirteen passengers and two crew members. The Hueys carry the same number of passengers and crew and can reach speeds of up to 130 knots.

The Boeing MD 600N is Boeing's largest single-engine helicopter and was certified by the Federal Aviation Administration in May 1997. It features double center-opening doors on both sides and several quick-change configurations depending on the operation. It also features a 115-gallon (435l) crash-resistant fuel tank and can stay airborne for 3.7 hours. Its Allison 250-C47M engines generate 808 horsepower. In August 1997, the U.S. Immigration and Naturalization Service announced plans to buy forty-five MD 600Ns for patrol functions along U.S. borders.

Boeing began production of the MD 500E in 1984. The craft is capable of lifting up to 1 ton (908kg) in a hover under the power of a single Allison 250-C20R engine with 450 shaft horsepower. It can reach a top speed of 175 mph (281.5kph) and can be configured for different operations.

OPPOSITE: *This 1,500-liter fire bombing bucket is similar to the Bambi Bucket, which is carried about 28 feet (8.53m) below the helicopter to drop water onto a raging blaze.* **ABOVE:** *Fixed-wing aircraft are not as versatile as helicopters, but they play a vital role in fire suppression, logging thousands of hours during a given fire season.*

The sheriff's aviation division began contracting with the California Department of Forestry in 1990 to offer mutual aid in initial attack fire suppression. The contract has allowed the county department to fund the purchase of additional aircraft—especially fixed-wing aircraft—for airborne patrol services. During the fire season, the Huey chopper is based at the Prado Fire Camp. This chopper has been equipped with Bambi Buckets since 1990 to fight wildland brushfires. These buckets hang about 28 feet (8.5m) below the helicopter for single-drop operation. The sheriff's department typically replaces its fleet of helicopters every five to seven years.

Even as fire and law enforcement departments switch to the Bell 412 to equip themselves with the latest in rescue and fire suppression technology, Bell

Helicopter Textron has developed yet another generation of helicopter. The Bell 427 was tested in Mirabel, Quebec, in December 1997, with scheduled certification for December 1998. It's powered by two Pratt & Whitney Canada PW206D turboshaft engines and seats a single pilot and seven passengers, with a number of different seating capacities for tailored use. If outfitted for emergency medical missions, the chopper can carry up to two litters.

The Bell 427 carries 190 gallons (719l) of fuel in its tank and can reach a maximum speed of 135 knots and a cruising speed of 130 knots. It can stay in the air for about four hours without refueling.

Swiss Air Rescue

The Switzerland-based Rega/Swiss Air-Ambulance Ltd. is quickly becoming a European leader in rescue/ambulance helicopter services, with an emphasis on bringing the hospital to the victim rather than just rapid delivery of the patient to the medical facility.

Rega operates a line of choppers that allow the administration of first aid at the scene of an accident during a rescue mission and before a victim is transported to a hospital. Rega's Agusta A-109-K2 is a flying intensive-care unit that carries medical apparatus onboard. Agusta produces an extensive line of helicopters for civilian and military use.

It's a rare thing to see a bi-plane dropping water or fire retardant on a wildfire, but they come in handy in a pinch.

The Agusta A-109-K2 is equipped with an electrocardiogram (EKG) heart monitor with a defibrillator and external pacemaker, as well as apparatus to monitor the heartbeat, blood pressure, and body temperature of a victim. Syringe pumps for infusions, a suction pump, a portable oxygen cylinder, and a physician's kit containing diagnostic equipment, infusion and intubation kits, drugs, and dressing complete the package. Each chopper also features a stretcher, scoop stretcher, and vacuum mattress.

The A-109-K2 is powered by twin French Turbomeca Arriel 1K2s, allowing the chopper an operational altitude up to 14,764 feet (4,500m) above sea level. It also features a rescue winch for two people with 164 feet (50m) of cable.

The United States Park Police use this Bell 412 Long Ranger for medical evacuation and transport. The interior of these Bells can be configured to accommodate any operation.

Russian MI-14 Amphibious Helicopter

Recent arrivals to the fire suppression scene are demilitarized Mi-14 amphibious choppers that have been converted into forest fire-fighting HeliTankers.

These new birds have been reconditioned specifically to fight brushfires with a tank system that carries more than 1,057 gallons (4,000l) of water. The chopper's two- and four-door drop system allows the pilot to determine which type of water or retardant drop to perform in the most effective manner. By having a water source within 6 miles (10km) of a fire scene, the Mi-14 can deliver up to 10,570 gallons (40,000l) of water each hour.

Mi-14s were manufactured by Mil Kasan in Russia between 1985 and 1988. A number of these choppers were reconditioned and sold to fire departments with less than twelve hundred hours on each airframe. Each is powered by two Russian TV3-117 turbine engines with a fuselage length of 59.74 feet (18m) and a width of 17.23 feet (5.2m) with the floats inflated. Weighing a maximum gross of 30,800 pounds (13,983kg), the craft also features a foam tank with a 105-gallon (397.5l) capacity. Its fuel tank holds 1,002 gallons (3,792.5l).

Fixed-Wing Air Tankers

While not as versatile as helicopters, fixed-wing air tankers play a vital role in fire suppression, logging thousands of hours during a given fire season.

The Hemet-Ryan Air Attack Base in southern California is a hotbed of activity between May and October, as the California Department of Forestry (CDF) battles blazes that often last days, sometimes a week. The total base landings at Hemet-Ryan for 1996 was 1,680. The total hours flown for CDF flyers was 2,777, and the Hemet-Ryan base accounted for 1,617 of those hours. It also used 2,183,577 gallons (8,265kl) of fire-retardant foam in its fixed-wing craft to fight fires. Costs to CDF to operate its fixed-wing program totaled more than $1.255 million. In all, 755 missions were flown from the base.

Fighting brushfires and forest fires is a dirty and dangerous business. On August 13, 1994, three firefighters were killed while responding to a massive blaze in Kern County, California. A C-130–Tanker 82–was piloted by Bob Buck. His copilot was Joe Johnson and the flight engineer was Shawn Zaremba. The three men were on a routine mission from Hemet-Ryan to Kern County when the C-130 went down.

A Hawaii Mars, foreground, and Philippine Mars at anchor, Sproat Lake, on Vancouver Island, British Columbia.

▲ ▲ ▲ ▲ ▲ ▲ ▲ ▲ ▲ ▲

Investigators believe the cause of the crash may have been an explosion in an auxiliary fuel tank. Twenty-four firefighters were killed between 1990 and 1997 while battling fires in the U.S. That's up four deaths from the 1980s.

While CDF owns and operates its own fleet of tankers, dozens of air tankers are owned and operated by private companies that contract their services to local, state, and federal agencies to fight forest fires. Aero Union Corp., Neptune Inc., Minden Air Inc., TBM Inc., Erickson Air Crane, J. Hirth Airtankers, the Hemet Valley Flying Service, and Hawkins & Powers are just a few of the firms that provide such services.

At the Porterville Air Attack Base in California, Aero Union, based in Chico, California, owns a Lockheed P2 Neptune that carries 2,000 gallons (7,570l) of fire retardant. Another tanker, similar to the old DC-4, is the C-54 air tanker, which also holds 2,000 gallons (7,570l) of retardant and uses a multidoor drop system. The plane is operated by Aero Flite of California.

The state of Minnesota typically uses PB4Y2s, P2-Ys, and P-3 Orions to fight fires out of its Brainerd air tanker base. These aircraft carry between 1,800 and 3,000 gallons (6,813 and 11,355l) of a mixture of water, clay, fertilizer, and color additive, which make up the fire retardant. These tankers can reload in less than ten minutes.

The PBY-1 and PBY-2 can typically be converted for fire suppression use. The PBY-2, for example, was designed initially as a patrol bomber that can seat a crew of nine. It's powered by a pair of Pratt & Whitney R-1830-64 engines to generate 900 horsepower. With a wingspan of 104 feet (31.5m), it measures 65 feet 10 inches (20m) in length and is rated at 28,400 pounds (12,893.5kg) gross weight. Its range is 2,110 miles (3,395km), with a cruising speed of 103 mph (165.5kph) and a top speed of 178 mph (286.5kph). Its ceiling is 20,800 feet (6,340m).

Air tankers are generally classified into four categories. The largest tankers are classified as Type I, including the C-130, P-3 Orion, and DC-7. These tankers carry about 3,000 gallons (11,355l) of fire retardant and are equipped with up to eight doors. Cruising speeds range from 235 to 275 knots. Type II air tankers include the DC-6, the PB4Y2, and both the DC-4 and DC-4 Super. Type II tankers carry between 1,800 and 2,999 gallons (6,813 and 11,351l) of retardant and are outfitted with up to eight doors. A DC-4 can achieve a cruising speed of 178 knots, but the PB4Y2 can hit 184 knots, the DC-6 about 215 knots.

A Hawaii Mars drops its 1200 gallons (4,542l) of water during a wildfire.

The B-26 bomber converted to fire suppression service carries 1,200 gallons (4,542l) as a Type III tanker, with a cruising speed of 200 knots and up to six doors. The smallest tankers, classified as Type IV, carry 100 to 599 gallons (378.5 to 2,267l) of retardant and include the Turbo Thrush, with a 350-gallon (1,325l) capacity and a cruising speed of 140 knots, and the Dromadear, which holds 400 gallons (1,514l) and has a 110-knot cruising speed. These tankers are equipped with up to two doors.

Since the end of World War II, the use of helicopters and fixed-wing aircraft to fight wildland fires has grown into a sophisticated operation. The Korean and Vietnam wars brought the technology, and U.S. Army and Air Force veterans brought the experience of flying these aircraft under the most arduous conditions.

Since 1950, 128 pilots, engineers, and firefighters in fixed-wing air tankers and helicopters have been killed in crashes while battling fires in the United States. It's a brutal job, but untold lives have been saved by the continuing advances in ways to attack fires by air.

FIRE
FIRE

6

WILDLAND

Previous pages: *Flames roll in dangerously close to this wildland fire unit in California. Departments responding to brushfires often require support from the air to protect their crews on the ground.* **Right:** *Hand crews are a vital part of the operation in battling wildland fires as they clear breaks and light backfires to control blazes.* **Above:** *This old but rugged 1974 AM General is a 6×6 with a 1500-gallon (5,677l) water tank and 250 gpm (946lpm) pumper. It serves Olympia, Washington.*

▲ ▲ ▲ ▲

Two fires nine years apart in the tough, hilly terrain of California would have a lasting impact on the way county and city fire departments battle wildland blazes.

In September 1923, downed power lines in the city of Berkeley started a fire that blackened 130 acres (52ha) and damaged or destroyed 484 structures. No one was killed, but it was the worst recorded fire during the early part of the century. The Matilija fire in Ventura County in September 1932 would destroy more vegetation–220,000 acres (88,000ha)–but it damaged no structures and claimed no lives.

It was these two blazes, and several smaller fires during the first three decades of the twentieth century, that drove home the reality that to meet the challenges of a major conflagration in an impossibly tough environment, it was critical that equipment be ready to meet the task.

The Moreland Truck Company of Burbank, California, was an early pioneer in manufacturing off-road vehicles exclusively for firefighting. In 1927, the Mount Shasta, California, fire agency took delivery of a single brush truck from Moreland. Three years later, the Los Angeles County Fire Department purchased several Moreland brush trucks,

The British Columbia Forest Service near Parksville operates this 1995 FL70 Anderson with a 500-gallon (1,893l) water tank and 30-gallon (114l) foam capacity.

▲ ▲ ▲ ▲ ▲ ▲ ▲ ▲ ▲ ▲

each equipped with a 650-gallon (2,460l) water tank and powered by a six-cylinder flathead Continental engine to fight wildland blazes.

California would continue to be assaulted by some of the most horrific wildland firestorms in the nation's history. And in each instance, firefighters took on a trench-warfare mentality with more rugged and more sophisticated equipment.

Three modern fires bring to mind the progression of state-of-the-art firefighting technology in terms of the vehicles used. The November 1980 arson-sparked Panorama fire in San Bernardino County, California, scorched 23,600 acres (9,440ha) and burned to the ground 325 structures, most of them homes. And two other arson-sparked fires erupted just days apart: the Laguna fire in Orange County, California, struck in October 1993, burning 14,425 acres (5,770ha) and destroying 441 homes and buildings; just days later, the unrelated Topanga fire in Los Angeles County blackened 18,000 acres (7,200ha) and burned 323 structures.

These are not isolated fires. The California Department of Forestry reports that on a five-year average, 160,584 acres (64,233.5ha) are burned annually, with 7,783 separate blazes reported. In 1997, 232,812 acres (93,125ha) were burned, with 8,134 separate fires reported. Most of these fires, however, are not the result of arson, but of accidents or—more often—natural phenomena.

Idaho has challenges equal to California's. In 1996, the Federal Bureau of Land Management (BLM) reported that 236,198 acres (94,479ha) were burned in Idaho's Shoshone and Burley Fire Districts; fifty-three of those 147 fires were caused by lightning. In the Idaho Falls Fire District, 206,888 acres (82,755ha) were burned, with fifty separate blazes ignited by lightning. For all of 1996, BLM reported that 445 separate fires burned 630,162 acres (252,065ha), a dramatic leap from 1994, when 355 fires blackened 168,675 acres (67,470ha).

Air support in the past thirty years has grown to high-tech levels, but it's the grunts on the ground that make the difference as to whether lives and property can be saved.

This 1994 Mack is operated by the California Department of Forestry. It has a 500 gpm (1,893l) pump and 650-gallon (2,460l) water tank.

▲ ▲ ▲ ▲ ▲ ▲ ▲ ▲ ▲ ▲ ▲ ▲ ▲ ▲ ▲

Off-Road Firefighting

Perhaps the most unique wildland vehicle to be delivered to fire agencies during the 1990s is the KME extended-cab Hummer Wildland Attacker, a brutish taskmaster sitting on a 130-inch (330cm) wheelbase with an overall length of 188 inches (477.5cm). It has a hefty ground clearance of 16 inches (40.5cm) and a 72-inch (183cm) track width. The Hummer has been used by the U.S. military for years and has recently gained popularity in civilian quarters. It is expected to prove its mettle battling fires.

Powered by a 6.5-liter V8 diesel engine generating 170 horsepower, the Hummer Wildland Attacker features a 200-gallon (757l) water tank and a Hale 20FD-L53 diesel pump with auxiliary fuel tank. It also features a polished aluminum booster reel with 200 feet (61m) of 1-inch (2.5cm) hose, capacity for 300 feet (91.5m) of $1^{1}/_{2}$-inch (4cm) hose, and a custom warning light package and rear deck lights.

Fire departments have the option of ordering the two-door model with a tank that holds up to 300 gallons (1,135.5l) of water, or the four-door model with a 200-gallon (757l) tank.

Pierce Manufacturing Inc. has come up with a four-wheel-drive Hawk Wildland

A hose/pumper truck, this one a converted Hummer, helps firefighters water down areas to slow an oncoming blaze.

Rapid Response Vehicle that moves firefighters over difficult terrain that larger, two-wheel-drive pumpers can't handle. The truck is designed to handle not only wildland firefighting but urban-interface fires, in which cities that abut rural areas are threatened by wildland blazes. As urban development encroaches into rural areas, a pumper that performs multiple duties on both the street and inhospitable terrain is critical for effective attack.

Right: *A fire truck carries firefighting personnel out of densely wooded areas.* **Above:** *This 1992 4×4 Hummer-Pacific has a 150 gpm (568l) pump, 330-gallon (1,249l) water pump and foam capabilities. The Hummer's 130-inch (330.2cm) wheelbase and hefty ground clearance of 16 inches (40.6cm) allow it to go almost anywhere.*

▲ ▲ ▲ ▲ ▲

Fighting wildfires is treacherous, grueling work not for the faint of heart. It is perhaps the most unpredictable and dangerous work of any firefighter, involving well-coordinated efforts from multiple teams on the ground and in the air.

The Wildland Rapid Response Vehicle is considered a first-strike, 33,000-pound (14,982kg) truck that tackles blazes in rural areas before homes are threatened. It's equipped with a 120-cubic-feet-per-minute (3.5 cu m) compressed-air foam system and an 850-gallon (3,217l) water tank. It can be equipped with foam tanks that give the truck a fire suppression capability of about 6,000 gallons(22,710l) of water.

The cab fits two firefighters and allows them to control a 300 gpm (1,135.5lpm) pumper turret that can discharge water, compressed-air foam, or a foam solution. It can also feature booster hose reels, cross-lay hose beds, and a single-stage centrifugal cross-mount water pump operated by diesel power. These trucks ride on Navistar Model 4800 four-wheel-drive chassis and are powered by diesel engines with horsepower ranging from 250 to 500.

S&S Fire Apparatus has introduced a similar interface vehicle dubbed the Outback. Considered a quick-attack vehicle for pumper, rescue, and brush work, the Outback sits on a Freightliner FL-70 four-wheel-drive chassis with a gross vehicle weight rating of 25,000 pounds (11,350kg). It features a 750-gallon (2,839l) water tank, capacity for 1,200 feet (366m) of 2 1/2-inch (6.5cm) hose, and 150 cubic feet (4 cu m) of compartment storage area.

E-One's interface vehicle is the Wildcat. Constructed on a Navistar 4900 four-door chassis with a 27,500-pound (12,485kg) gross vehicle weight rating and 176-inch (447cm) wheelbase, the truck has a pair of 25-gallon (94.5l) foam tanks and a 500-gallon (1,892.5l) water tank. Its booster reel provides for 200 feet (61m) of 1-inch (2.5cm) hose and a Foam Pro 2001 foam system. Power under the hood comes from an International 530 E 300-horsepower engine with an Allison MD 3560P transmission.

E-One's 35,000-pound (15,890kg) Grizzly elliptical tanker is one of the largest wildland tankers available, with a Ford F-800 chassis and power from a 250-horsepower Ford diesel engine. Behind the cab is a 500 gpm (1,892.5lpm) pump and a 2,000-gallon (7,570l) aluminum tank, plus 24 cubic feet (0.7 cu m) of compartment storage.

The Colorado Fire District has an extensive fleet of military-surplus wildfire engines that are spread throughout the state. More than 150 wildfire engines are at the ready to respond to any emergency.

The California Department of Forestry uses this 1994 International-Harvester 4900 with a 500-gallon (1,893l) water tank.

Sixteen engines are assigned to the volunteer fire departments in the Grand Junction district. All repair and maintenance of the trucks is performed by the fire district's shop in Fort Collins.

Virtually all of Colorado's trucks are 4×4 and 6×6 all-wheel-drive machines equipped with water tanks ranging from 200 to 1,000 gallons (757 to 3,785l). They also come with pumps, hose reels, and equipment compartments. Much of the older equipment is updated with foam-injection systems, while new surplus models are equipped with compressed-air foam systems, making water distribution at fire scenes ten times more effective.

Trucks carrying compressed-air foam systems are mounted on 5-ton (4.5t) 6×6 chassis. Many other wildland engines are built on 3/4-ton (681kg) military chassis, with the 1,000-gallon-water-tank (3,785l) engines mounted on a 2 1/2-ton (2.3t) 6×6 military chassis.

Fighting wildfires is perhaps the most unpredictable and grueling job for today's firefighter–and probably the most exciting. Firefighters are faced not only with nature and its wildlife and unfamiliar geography but often with nearby urban developments in need of protection. It means a two-front war: a war from the ground as well as the air.

7

WATERCRAFT

During the winter of 1998, the foothill communities in southern California were awash in rivers of rainwater tumbling from mountain areas into washes and flood control channels.

Inevitably, youngsters, young adults, and wayward motorists found themselves trapped in the raging torrents.

In one instance in Ontario, California, a young man and the firefighter attempting to rescue him were swept 11 miles (17.5km) down a flood control channel as El Niño dumped 3 inches (7.5cm) of rain in a single afternoon. In the city of Los Angeles alone there are 470 miles (756km) of flood control channels and more than 2,000 miles of storm drains.

PREVIOUS PAGES: *Historical buildings and fine architecture often found on waterfronts need constant protection from both land-based fire suppression units and watercraft. It's a comforting feeling for building owners and city leaders to know that watercraft can sidle up to a building at a moment's notice.* **LEFT:** *Seattle (Washington) firefighters battle a fire on a cargo ship. Such firefighters are faced with many problems not encountered in land-based blazes, including steep ladders and slippery steel surfaces, extreme heat due to steel construction, high voltage areas and blind spots inside the craft.*

Water rescue units, such as the Lake George (New York) Volunteer Fire Department Scuba Rescue truck, are vital for responding to emergencies in communities with recreational water areas. Even in municipalities with modest budgets and volunteer fire services, city and town leaders recognize the necessity of such equipment.

▲ ▲ ▲ ▲ ▲ ▲ ▲ ▲ ▲ ▲

Los Angeles city and county firefighters and rescue teams must contend with an average of one hundred channel incidents and as many as six drownings a year in such events.

In the coastal areas, water rescue and the role of the firefighter–and, by extension, that of the lifeguard–are entirely different. Water rescues performed by the San Diego (California) Lifeguard Service numbered 7,236 in 1996. There were 137 rescues of boats and thirty-six boat fires and pump-outs of sinking vessels during 1996.

At the Los Angeles and Long Beach harbors, the busiest ports of entry on the West Coast to serve all of the Pacific Rim countries, nearly 30 miles (48km) of modern cement docks for containerized cargo, petroleum, and chemical shipments require precision coverage by the Los Angeles Fire Department's highly trained fireboat crews.

And in Boston, the U.S. Coast Guard must contend with sinking barges that have fuel tanks loaded with 400 gallons (1,514l) of diesel fuel, or fishing vessels taking on water in high seas.

All of these incidents require sophisticated watercraft equipment, whether it's a swift-water boat, a 100-foot (30.5m) fire pumper with massive 1 gpm (3.7lpm) pumping capacity, or a scuba team fighting a blaze below water and under a wharf. Every major harbor has at least one fireboat to protect docks and ships where land-based firefighting vehicles can't go. Fireboats originated in the late nineteenth century, when tugboats were converted for fire duty. By the early twentieth century, these boats had proven their worth, and cities began to commission boat designs specifically for firefighting and rescue.

Marine firefighting for the land-based firefighter is an exhausting, dangerous duty, fraught with complications not seen in typical land firefighting. Cities such as Seattle, Boston, Los Angeles, San Francisco, New York, and Miami–and to a lesser degree Chicago and Detroit–have firefighters trained in marine firefighting. These crews must respond to blazes aboard or associated with waterborne vessels. These vessels might be found at a dock or wharf, but also at anchor several miles from shore; the fire may be on a small runabout

Local police agencies must equip themselves with light watercraft for small bodies of water.

or a supertanker. This type of firefighting often involves haz-mat responses because of the contents of a cargo ship or the fuel it's carrying.

Fighting fires aboard ships is often complicated by disputes over which agency has ultimate authority over the firefighting operations. The master of a ship may simply refuse to allow firefighters to board his vessel, but he could be overruled by the Port Authority or the Coast Guard. Though the fire department may maintain complete control over the firefighting operations, it may have to answer to a number of different agencies during the course of the operation.

Once onboard a ship, firefighters are faced with a variety of problems not encountered in a standard land procedure. Potential shipboard hazards include steep ladders, enormous heights, slippery steel surfaces, extreme heat due to steel construction, excessive steam generation, high-voltage areas, an inordinate number of blind spots inside the ship, and even the movement of the ship itself. In addition, non-English-speaking (in the United States) and nonuniformed ship personnel may delay or hinder the quick response to a fire call aboard a vessel.

Firefighting techniques aboard a ship are much different from those of land-based fires. Firefighters can close all ventilation

A 1-ton (.907t) Ford utility truck, a light boat attached to the top and a trailered watercraft behind provide this dive-water rescue team with all the essentials for handling most emergencies on a lake or river. **Opposite:** *The Anthony J. Celebrezze fireboat patrols the waterfront for the Cleveland (Ohio) Fire Department.*

openings on a ship and shut off fuel supplies to effectively place a lid on the blaze. Cooling down the exterior of a vessel also allows an interior blaze to burn out safely. Tools used to put out vessel fires are foam, Halon, high-pressure fog, high-expansion foam, dry chemicals, sprinklers, and a deluge system. Because of environmental concerns, fire departments rarely, if ever, tow a ship out to sea and sink it.

The Seattle Fire Department has one boat on duty at all times to protect its harbor and docks. In May 1997, dozens of firefighters responded to a blaze beneath Ray's Boathouse at Shilshole Bay. The blaze burned for almost six hours as the department's land-based fire engines and fireboat wrestled for control. Two months later, another fire erupted at a deli on Pier 70, causing an estimated $6,000 in damage. It could have been much worse it if hadn't been for Seattle's fireboat navigating under the pier to spray a steady stream of water through the pier's floor.

Fireboats

A typical fireboat is powered by twin Detroit Diesel 8V92TA engines that generate 600 horsepower at 2300 rpm. The Detroit Diesels are freshwater-cooled with wet exhausts and fiberglass mufflers. Power is transmitted through twin disc MG 514-M Omega Marine gears with a 1.5:1 ratio. This type of fireboat carries 300 gallons (1,135.5l) of fuel and a water-pumping capacity of 5,000 gpm (18,925lpm) at 150 pounds per square inch (psi). Such boats are also equipped with 24-mile-range (38.5km) radar and digital low-range depth finders. Prices for these fireboats start at about $750,000.

The New York Fire Department has one fireboat that is still active after sixty years of service. The *Firefighter* began service in 1938. It was designed by William Francis Gibbs, who also designed the *America* and *United States* ocean liners.

The *Firefighter* features four 5,000 gpm (18,925lpm) De Laval two-stage centrifugal pumps with a total pumping capacity of 20,000 gpm (75,700lpm) at 150 psi. The pumps can be connected to another system that provides 10,000 gpm (37,850lpm) at 300 psi.

Much of the technology used by today's fire departments to combat wharf fires was developed by the LAFD in the early 1960s. The department developed a fleet of fireboats, scuba gear, and special tools to cope with the unique needs of harbor firefighting.

The department today uses five fireboats. The *Ralph J. Scott* is the oldest and

largest at 100 feet (30.5m) long with a pumping capacity of more than 1,700 gpm (6,434.5lpm). It features five turrets to house fire hoses and protect firefighters. The turrets are found in the tower and pilot house, and at the stern and bow. The *Ralph J. Scott* entered service in 1925 and is docked at Fire Station No. 112. The fire station is a two-story facility with 17,823 square feet (1,657.5 sq m) to house the fireboat, a rescue ambulance, and two land-based firefighting apparatus.

The department's *Bethel Gifford* is 74 feet (22.5m) long with a pumping capacity of 2,000 gpm (7,570lpm). Los Angeles' remaining three boats are each 34 feet (10m) long with 750 gpm (2,839lpm) pumping capacities and are used primarily as scuba operations platforms. Each is manned by a fireboat mate and two firefighter/scuba divers. Working independently or as a team with the other boats or land-based fire engines, the small boats use direct-attack strategy or serve as support to the other vehicles. They function best dealing with boat fires, wharf

Fighting pier or wharf fires is a tricky business involving many hidden blazes and difficult access to underwater fires in air pockets. **Opposite:** *This craft is used primarily for patrol, rescue, and body retrieval.*

and underwharf firefighting, drownings, haz-mat spills, inspections of petroleum tankers, helicopter support, underwater search and rescue, boat rescue, and dewatering partially sunken vessels. Each of the 34-foot (10.5m) boats is equipped with a mounted 2 1/2-inch (6.5cm) bow turret and 1 1/2- and 2 1/2-inch (4 and 6.5cm) fire hoses, as well as breathing apparatus, underwharf firefighting floats, standard entry tools, and a pressure tank containing 50 gallons (189l) of foam.

Vancouver (British Columbia) Fire and Rescue Services has a floating fire suppression fleet that rivals that of Los Angeles and surpasses Seattle's. At five separate locations–two in the city of Vancouver and one each in North Vancouver, Burnaby, and the District of North Vancouver–fireboats built in 1992 provide 3,039 gpm (11,500lpm) in pumping capacity for a variety of fire calls to waterfront areas and the region's vast waterways.

An impossible job: working in the darkness to rescue near-drowning victims or stranded boaters. A scuba diver's job is a dangerous one. **Opposite:** *Foul weather doesn't deter the rescue team from saving the lives of boaters.*

▲ ▲ ▲ ▲ ▲ ▲ ▲ ▲ ▲ ▲

Swift-Water Rescue

Imagine a fourteen-year-old boy who thought he could ford a rain-swollen flood control channel only to be swept through three counties on his way to the Pacific Ocean. It happens. And it happens a lot during the rainy season in California.

Swift-water rescue is probably closer to an exact science than any other firefighting practice or rescue performed by fire department personnel. Not only does it take coordination and timing to capture a boy floating down a channel at 10 knots, but it takes the cooperation of surrounding agencies and jurisdictions. Without the help of perhaps half a dozen fire departments on a particularly dramatic rescue, that boy could be swept out to sea, never to be heard from again.

Many small and medium-size fire departments have variations of swift-water rescue teams, but the LAFD is one of the leaders in the country, or at least on the West Coast. All LAFD companies are trained and equipped for land-based river rescue, and all carry flotation rings, helmets, personal floatation devices, and throwbacks. The specialized swift-water rescue team performs the actual water rescue whenever possible.

Responding to a water rescue incident is not a simple matter. Waterways must be mapped and variable water speeds must be calculated. Firefighters must pinpoint the victim's rate of travel and establish rescue sites at specific areas based on these calculations. A swift-water rescue team is elaborate. It consists of one assistant fire chief, a battalion chief, two task forces that include a hook-and-ladder truck and a two-piece engine company, three single-engine companies, one rescue ambulance, a pair of helicopters, and two rescue teams each consisting of four firefighters and two Los Angeles County lifeguards. In all, that's fifty-seven emergency personnel to save a single person from drowning.

Land-based rescues are generally attempted first by using throw bags or flotation rings, as well as suspending an air-inflated 2 1/2-inch (6.5cm) fire hose across a channel, stream, or river to give the victim something to hang on to. The swift-water rescue team, generally using a Yamaha personal watercraft, performs the in-water rescue using a rescue board, nets, or ropes.

Lifeguard Vessels

The San Diego Lifeguard Service operates a pair of 32-foot (9.5m) fire/rescue watercraft. Each craft has the capacity to pump about 1,000 gallons (3,785l) of sea water per minute. These craft are also capable of pumping as much as 300 gallons (1,135.5l) of water per minute out of sinking vessels. Each comes equipped with Automatic Direction Finders and a Global Positioning Satellite to track distressed vessels, especially in heavy fog or poor weather conditions.

For surf rescue, San Diego equips its lifeguards with 22-foot (6.5m) Boston Whalers that are modified to deflect oncoming waves. The Boston Whalers are powered by 250-horsepower outboard engines that can achieve 40 mph (64.5kph) with a full load of medical and underwater search equipment.

While not necessarily used by the San Diego Lifeguard Service, a watercraft such as the Oceanid Fortuna is a prime example of the type of equipment used by fire departments, lifeguards, and law enforcement agencies.

The Fortuna is an unusual-looking nonmotorized craft that has extremely rockered ends. These upturned ends allow rescuers to maneuver the boat's open end over the victim while the victim's head is still above water. The craft's freeboard height remains low to allow the rescuer to pull a victim into the boat with minimal effort.

For body retrieval, the body can be slipped through the access portal to be brought to shore. The Fortuna ferries similar to a canoe and can be controlled by one rescuer while the other saves the victim. The Fortuna can be stored in a 2-foot (61cm) cube, carried by one person, and inflated within seconds.

Oceanid's PowerCats are motorized units that can travel anywhere at high and low speeds and move in as little as 6 inches (15cm) of water.

As the role of the firefighter continues to change, especially in a society that treasures recreational activities, watercraft play an increasingly important role in saving lives and fighting fires.

8

Hazardous Materials and Urban Rescue

Previous pages: *This haz-mat crew is wearing Level A protection, which includes a self-contained breathing apparatus (SCBA); positive-pressure, supplied air respirator; an encapsulating chemical protective suit; outer and inner chemical resistant gloves; chemical-resistant boots with steel toe and shank; and a two-way communication system.* **Right:** *A haz-mat crew uses Level A protective clothing to respond to an accident involving a passenger car that collided with a train pulling a tanker with hazardous chemicals.* **Above:** *The Creve Coeur (Missouri) Fire Department uses this 1996 Pierce 4×4 heavy rescue unit.*

▲ ▲ ▲ ▲

The twenty-first century sparks the beginning of a new wave of changes for today's American and European fire departments.

The roots of these developments are found in the past twenty-five years, as fire departments diversified services to communities.

These changes find fire suppression in urban areas taking a back seat to medical emergency responses and hazardous-materials recovery.

The new millennium promises that the current trend in fire departments will focus on haz-mat recovery, and a new emphasis will be placed on a relatively new technique: urban search and rescue. Haz-mat operations developed in the early and mid-1970s, not because of communities' concerns over exposure to toxic chemicals and the proliferation of illegal drug labs but because of new federal regulations that put the responsibility of hazardous-materials recovery squarely on the shoulders of local fire departments. Local fire agencies perfected the quick response. With more sophisticated and strict fire safety regulations enacted since the mid-1960s, it was a logical choice for fire departments to assume the new role of acting as society's cleanup crew for late-twentieth-century messes.

Urban search and rescue operations are an even newer development brought on by domestic terrorism in the 1980s and 1990s and by lessons learned from a series of devastating earthquakes ranging from the Mexico City quake of 1985 to the Loma Prieta quake in the San Francisco-Oakland area in 1989. If local fire departments learned anything from these earthquakes and from the bombings of New York City's World Trade Center in 1993 and Oklahoma City's Federal Building in 1995, it is that traditional fire department operations are insufficient for dealing with large-scale disasters. As the world community becomes smaller in the twenty-first century, with demands on U.S. and European fire departments to respond to emergencies worldwide, the urban search and rescue and hazardous-materials crews will become nearly as vital as the United Nations peacekeeping forces.

Haz-Mat

Hazardous-materials recovery was a long time in coming. It wasn't until the 1970s that the federal government stepped in and delegated authority for the disposal of toxic chemicals. At the time, there were few laws regulating the transportation, storage, and disposal of hazardous chemicals. Local agencies in particular were reluctant to assume the liability if hazardous materials were spilled in urban areas.

Still, communities were faced with more instances of the public being exposed to chemical dangers. As residential development encroached on industrial areas and on land traditionally used for agricultural purposes, it became increasingly apparent that steps had to be taken to insure public safety. In addition, the transportation of products used by the public yet extremely dangerous in raw form exacerbated the threat of danger. Sophisticated haz-mat operations remain crucial to public safety when one considers that in addition to the now-traditional haz-mat operation, the new threat of domestic terrorism is all too real, with such deadly bacteria as anthrax a possible weapon.

In 1976, Congress passed into law the Toxic Substances Control Act, the Resources Conservation and Recovery Act, and the Comprehensive Environmental Response Act. Local and state fire agencies were now compelled to develop haz-mat response teams to deal with the new responsibility.

Those early years saw fire departments tackling haz-mat operations using a trial-and-error method, which was costly in terms of personnel lost to injuries and lost equipment. It soon became apparent that traditional firefighting methods would not work in cleaning up toxic chemical spills. In the late 1970s and early 1980s, fire departments underwent radical changes to accommodate these new responsibilities, changes that ensured the safety of personnel and reduced the loss of equipment.

Some departments responded more quickly than others. After all, hazardous-materials recovery accounts for less than 5 percent–even 1 percent in smaller agencies–of emergency calls in most fire departments. In fact, a typical North American community can expect a major natural or man-made disaster once every twelve years. But those disasters and subsequent emer-

gency calls are costly and dangerous to firefighting personnel. In the 1950s and early 1960s, less than 5 percent of the emergency calls were for medical services. That figure has climbed to an average of 70 percent in thirty-five years. Clearly the need for hazmat operations will rise in the next century.

There are many different ways a firefighter can be exposed to hazardous chemicals: toxins can be inhaled, ingested, absorbed, or injected during many kinds of rescue operations. Personnel can be exposed to toxins from clandestine drug labs and in the form of exhaust fumes, cleaning supplies, pesticides, and herbi-

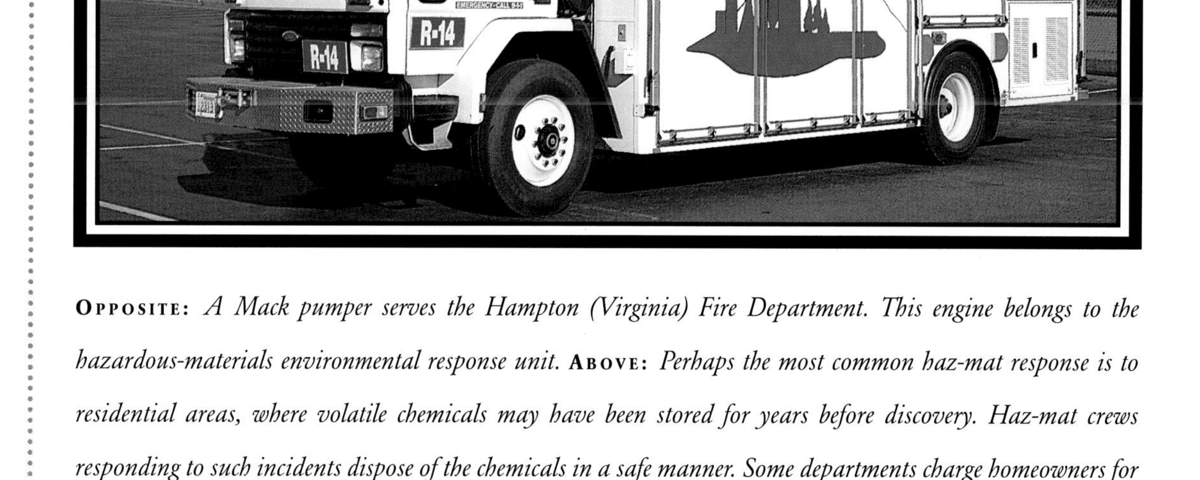

Opposite: *A Mack pumper serves the Hampton (Virginia) Fire Department. This engine belongs to the hazardous-materials environmental response unit.* **Above:** *Perhaps the most common haz-mat response is to residential areas, where volatile chemicals may have been stored for years before discovery. Haz-mat crews responding to such incidents dispose of the chemicals in a safe manner. Some departments charge homeowners for the service if negligence is determined.* **Inset:** *Difficult rescues in Seattle, Washington, are performed by the city's Technical Rescue unit, which is equipped with this 1995 Ford Cargo Mobile Heavy Rescue apparatus.*

A haz-mat crew attempts to identify dangerous chemicals during an operation.

cides. Firefighters can also be exposed to toxins from combustion at structure, vehicle, and vegetation fires. During medical emergencies, crews may be exposed to communicable disease.

Exposure to these dangers demands that firefighters be adequately protected. Federal guidelines rate protective clothing from Level A, which is the most protective clothing, to Level D, which provides minimal protection.

The highest level of protection, Level A includes self-contained breathing apparatus (SCBA), a positive-pressure, supplied-air respirator with escape for a five-minute minimum duration; an encapsulating chemical protective suit; outer and inner chemical-resistant gloves; chemical-resistant boots with steel toes and shanks; and a two-way communication system.

Level B is the same as Level A, but a two-way communication system is not required. However, most fire departments make little distinction between the two levels and demand continuous communication with the command center and the line firefighter.

Level C requires a full- or half-face mask, air-purifying respirator, and chemical-resistant clothing, gloves, and boots.

For Level D, respiratory protection is not required and coveralls can replace hooded chemical-resistant clothing. Chemical-resistant boots with steel toes and shanks are required.

U.S. fire departments, especially in cities with populations of more than 100,000 people, can receive federal grants to outfit their departments with state-of-the art haz-mat equipment. The Riverside (California) Fire Department operates a motorhome that can be transformed into a mobile/office command center so that the firefighters can monitor a haz-mat incident and direct operations from a safe distance.

The command center is part of a seven-man crew with three motor units that include a first-responder truck and an auxiliary truck, which provides additional equipment, including additional air tanks. The command center is equipped with a four-band radio, a television to monitor news reports of a major incident, and a video camera to record a scene for study and to monitor wind conditions that may impact a chemical hazard.

The mobile center also contains an elaborate computer system that allows personnel to enter data gleaned from the scene to help identify chemicals. With the computer, firefighters can instantly identify the structure and properties of specific toxins, then contact chemical companies via fax machine to determine the level of danger.

Urban Search and Rescue

A box dripping with red liquid was discovered at the B'nai B'rith office in Washington, D.C., in 1997. Office workers called the fire department, which had an emergency response team open it. Inside was a petri dish labeled "anthrax." A year later, two men were arrested in Las Vegas carrying what was believed to be anthrax.

In both cases, the contents were harmless, but the unrelated incidents served as a wake-up call to the federal government. Ten pounds (4.5kg) of the powdery bacteria could kill as many as 300,000 people in a city of half a million.

The anthrax scares are part of an emerging pattern of incidents that paint a picture of domestic terrorism as an increasingly viable threat to the safety of the public. The bombing of New York's World Trade Center in 1993 and the bombing of the Alfred P. Murrah Federal Building in Oklahoma City in 1995 illustrate that domestic terrorism is now part of the fabric of our daily lives. There have been other incidents that also showed law enforcement authorities just how easy it is to strike fear in public places. A doomsday cult released sarin gas in a Tokyo subway

Above: *The South Davis (Utah) Fire Department employs this 1992 Pierce-Lance heavy rescue truck. Many small communities that are connected to railroads require such equipment.* **Right:** *The command center inside a motor home converted for haz-mat operations. The television at right allows crews to monitor news reports or use a video camera to safely survey the scene. Short wave radio allows crews to monitor communication traffic. The computer at left allows the team to feed information into a database that will help them identify the hazardous materials they have encountered.*

in 1995. Another cult spread salmonella at restaurants in Oregon to make people sick so that they couldn't vote in a local election. The same group had also planned to contaminate the local water supply.

These terrorist acts prompted six federal agencies to develop a domestic preparedness program. The Defense Department, the Health and Human Services Department, the Environmental Protection Agency, the Public Health Service, the Federal Emergency Management Agency (FEMA), and the FBI banded together to train fire and search and rescue personnel from 120 cities in 1997 and 1998. Training consists of "responder awareness," in which teams are instructed on basic terrorism and identification of methods and groups; "target hazards," which trains teams to identify a hazard; and what to do on the scene of a potential biological disaster. Haz-mat teams in particular have undergone intense training in how to respond to a chemical bombing incident.

Through 2000, the federal government will provide principal funding, advice,

Left: *A winch is used to help a rescue team pluck a woman from a deep wash.* **Opposite:** *A 1996 Ford with an International truck body serves a technical rescue unit for the city of North Vancouver in British Columbia.*

training, and limited equipment (detection devices and high-tech protective suits costing about $3,000 apiece) to protect what are deemed high-risk targets. Federal buildings, college campuses, and courthouses are considered by terrorists to be targets that carry the maximum public impact.

While these moves are designed to counter domestic terrorism, federal, state, and local agencies also consider natural disasters worldwide. Mexico City, Soviet Armenia, and California's Bay Area have been struck by catastrophic earthquakes since 1985. Since about 1978, three million people have been killed in worldwide isasters. In that same time, an estimated $50 billion in property damage has been caused, and about 800 million people have been adversely affected by these disasters. An estimated $1 billion per week has been spent by the U.S. government as a result of natural disasters in the past twenty years.

Urban search and rescue teams, in addition to responding to terrorists acts like the Oklahoma City bombing, respond to these types of urban disasters. The purpose of these teams is to locate and rescue people who are victims of natural and man-made disasters of a massive scale.

The urban search and rescue medical teams, as part of the FEMA task force system, are to local and state fire agencies in the 1990s what haz-mat operations were to these departments in the 1970s. And while these teams have reached sophisticated levels, the program was developed over twenty years based on the observations of dozens of natural and man-made disasters.

The federal government recognized that local fire agencies were severely taxed, if not overwhelmed, in the hours immediately following a disaster. Chaos at such disasters as a construction site collapse in Bridgeport, Connecticut, in 1987 and a department store collapse in Brownsville, Texas, in 1988 showed that local agencies are often ill equipped to supervise a massive operation. Lightly trapped survivors of such disasters are often rescued by bystanders who have no training in rescue technique or medical care. And victims trapped deeply in a

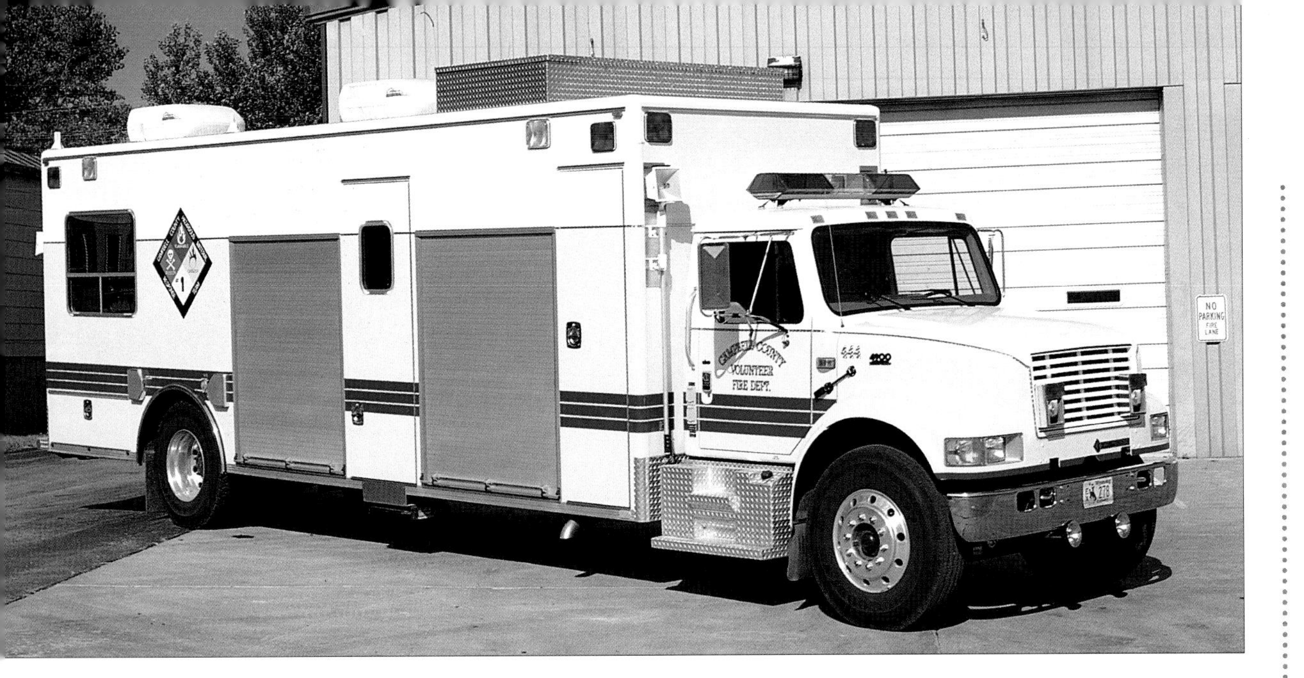

The Campbell County (Wyoming) Fire Department employs a 1995 International-Harvester 4900 Amtech for its haz-mat command center. It's also equipped with an air compressor to fill air bottles. Compartments on these units are as deep as 28 feet (8.53m).

▲ ▲ ▲ ▲ ▲ ▲ ▲ ▲ ▲ ▲

collapsed building are beyond detection or extrication by poorly equipped rescuers.

Each team consists of structural engineers; hazardous-materials specialists; heavy-rigging specialists; search specialists accompanied by trained search dogs; logistics, rescue, and medical specialists; and at least two physicians. Specific medical teams include four medical specialists and two doctors, accompanied by paramedics and firefighters. Each medical team is designed to bring the emergency department to the scene of a disaster.

The California Task Force 7 consists of the Sacramento Urban Search and Rescue Team and the Sacramento Fire Department, and can respond to any national emergency at a moment's notice. It was mobilized for the Northridge earthquake in 1994 and the Oklahoma City bombing a year later. In July 1996, it responded to the Yosemite rock slide, in which 30 tons (27t) of granite from Glacier Point broke loose, creating 300 mph (483kph) winds and knocking down more than three hundred trees over a quarter-mile (402m) area. The team responded to the incident and rescued dozens of people. Only six people were hurt and one was killed.

At the 1996 Olympic games in Atlanta, FEMA deployed a dozen urban search and rescue teams, including the California Task Force 7, as standbys for the games between July 25 and August 1. The teams were deployed when a bomb went off on the festival grounds.

Los Angeles County's Disaster Medical Assistance Team (DMAT) is also part of the mutual aid program through FEMA. Depending on the disaster, DMAT's team consists of about thirty-five members: four or five physicians, a dozen nurses and paramedics, as many as twelve emergency medical technicians, a communications specialist, a respiratory therapist, and a pharmacist. More than two hundred members make up the entire DMAT, so that two or more teams can be dispatched at any one time.

While DMAT responds through the federal mutual aid program, it can also be activated by county and state agencies and deployed to other counties. These teams can be deployed for up to two weeks at a time, and operate from a fixed facility or their own temporary quarters. Each team must be self-sufficient for up to seventy-two hours.

The Los Angeles County operation is typically supported by an SH-3H Sea King helicopter with a 14,700-foot (4,480.5m) ceiling height and a range of 542 nautical miles. The Sea King is powered by twin GE T58-GE-10 turboshaft engines and can achieve a maximum speed of 166 miles per hour (267kph).

Between August 1991–when FEMA organized its task force system–and January 1998, urban search and rescue teams have responded to thirty-four disasters. Imagine a Korean War–era M*A*S*H unit–a mobile

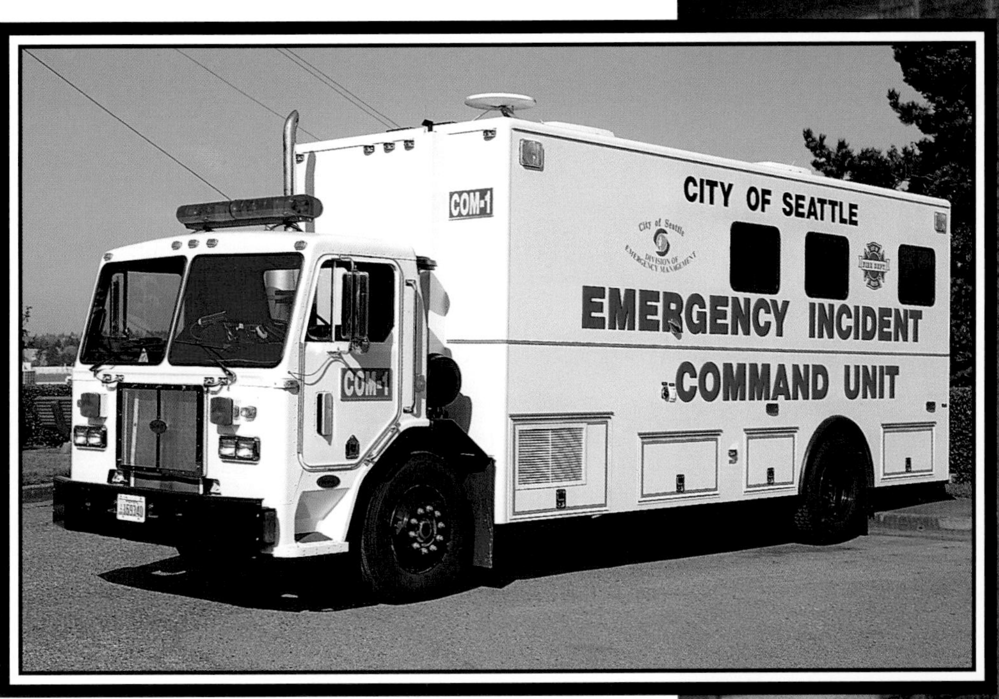

Above: *A 1993 Peterbilt mobile technical command center used by the Seattle (Washington) Fire Department.* **Right:** *Fire crews often can't rely on ladders or aerial platforms for rescue efforts in tall buildings. Here, firefighters use a steel cable and winch for a rescue operation.*

▲ ▲ ▲ ▲ ▲

hospital, both in the air and on the ground–with state-of-the art equipment, including mini-surgical units, quick transport, and skilled searchers aided by canine units to quickly locate and rescue survivors of a disaster.

Since the inception of the federal program in 1991, hundreds if not thousands of lives have been saved. Meanwhile, FEMA is expanding its operation worldwide to respond to not only natural and man-made disasters, but war-torn areas as well.

Selected Bibliography

Burgess-Wise, David. *Fire Engines and Firefighting.* Norwalk, Conn.: Longmeadow Press, 1977.

Cottrell, William H., Jr. *The Book of Fire.* Missoula, Mont.: Mountain Press Publishing, 1989.

Halberstadt, Hans. *The American Fire Engine.* Osceola, Wis.: Motorbooks International, 1993.

International City Management Association. *Managing Fire Services.* Washington, D.C.: ICMA Training Institute, 1988.

Photography Credits

©Richard Adelman: pp. 29, 31, 49 right, 51, 55, 67, 108 inset

AP/Wide World: p. 117 right

Arms Communications: ©Robert F. Dorr: p. 71; ©Douglas A. Zalud: p. 81

©Frank Degruchy: pp. 33, 40, 42 inset, 82, 83, 88, 115
©Dembinsky Photo Associates: pp. 6–7, 17, 27, 43, 98

©William B. Folsom: pp. 73, 76

©Bill Hattersley: pp. 2–3, 25, 26, 28, 30, 39, 44 bottom, 45, 48 inset, 50 inset, 52, 53 inset, 54 inset, 58 inset, 59, 61, 62, 66, 86 inset, 89, 91 top, 93, 111 inset, 113 top, 116, 117 inset

New England Stock Photo: ©Eric R. Berndt: p. 41; ©David G. Curran: p. 54 left; ©Fred M. Dole: pp. 3 right, 22–23, 32; ©David A. Frye: p. 36 inset; ©Leonard C. Lacue: p. 12; ©Pat Lynch: pp. 53 left, 70; ©David E. Rowley: pp. 84–85; ©Jim Schwabel: pp. 10–11

911 Pictures: ©Adam Alberti: p. 16; ©Bill Bongiorno: pp. 68–69, 99; ©Kari K. Brown: pp. 86–87; ©Bruce Cotler: p. 103; ©Dean Culver: p. 92; ©Pete Fisher: pp. 74, 105, 108–109; ©Jeremy Greene: p. 114; ©Michael Heller: pp. 46–47, 48–49, 56–57, 80, 90, 91 inset, 111 top, 113 inset; ©Mary Myers: p. 50 left; ©Chris E. Mickal: pp. 21, 38 left, 106–107; ©John Mielcarek: pp. 24, 34–35; ©Joaquim Pol: p. 112; ©Neil Schneider: p. 60 top; ©Steve Spak: pp. 20, 36–37, 58 right, 94–95, 102, 104; ©Karen Wattenmaker: pp. 18, 78, 79
©Bill Stormont: pp. 19, 38 inset

Unicorn Stock Photos: ©David Cummings: pp. 96–97; ©Jean Higgins: p. 44 inset

UPI/CORBIS-Bettmann: pp. 8–9, 14–15

©Douglas A. Zalud: pp. 5, 13, 42 left, 60 inset, 63, 64 both, 65, 72, 100, 101, 110

Zalud Collection: © Dr. J.G. Handelman: p. 77

Index